Library of
Davidson College

VOID

THE ECONOMICS
OF INVENTION
AND INNOVATION

by the same author

CO-OPERATIVE RESEARCH IN INDUSTRY

THE ECONOMICS OF INVENTION AND INNOVATION

With a Case Study of the Development of the Hovercraft

P S Johnson
Lecturer in Economics, Durham University

Martin Robertson

© P. S. Johnson 1975

All rights reserved. No part of this publication may be reproduced, stored in a retrieval system, or transmitted in any form or by any means, electronic, mechanical, photocopying, recording, or otherwise, without the prior written permission of the copyright holder.

First published in 1975 by Martin Robertson & Co. Ltd., 17 Quick Street, London N1 8HL.

ISBN 0 85520 078 2

338.9
J68e

76-9652

Set by Trade Linotype Ltd., Birmingham, reproduced and printed by photolithography at The Pitman Press, Bath.

Contents

	Preface	vii
	PART 1: Some Economic Issues	
1	Technological Change	3
2	The Process of Invention	29
3	Size of Firm, Invention and Innovation	51
4	Competition in Innovation	72
5	Scientists and Engineers : their Supply and Demand	92
	PART 2: Channels for Technological Change	
6	The Government's Role	115
7	Technological Change in Industry	156
8	Co-operative Industrial Research	190
9	The Universities' Contribution	209
	PART 3: A Case Study in Invention and Innovation: the Development of the Hovercraft	
10	The background to the Study	229
11	Hovercraft Production, Operations and Markets	236
12	The Individual Inventor	244
13	The Technical Development of the Air Cushion Principle	251
14	The Development of the Industry	265
15	Public Involvement	298
16	Conclusions on the Case Study	309
APPENDIX 1	Time Savings (to 1972) from Hovercraft Operation	317
APPENDIX 2	Total R & D, Production and Operating Costs (to 1972) of Hovercraft Development.	320
	Further Work	322
	Index	325

to KATHRYN

Preface

The initial stimulus for this book came from a realisation that there was no text that dealt reasonably comprehensively with the more applied economic aspects of technological change within the context of British experience, and which was suitable as an introductory text for undergraduates taking courses in the economics of industry and related subjects. I was also aware that the non-economist in education, industry and government who is concerned in one way or another with the process of technological changes would find it extremely difficult to gain a quick, overall impression of the contribution the economist has made to the understanding of this process.

Mansfield[1] and Nelson et al.[2] have provided very useful texts in this field, but the orientation in their treatment of institutional aspects is almost wholly towards the American economy. Although much of the material presented in this book is drawn from American sources, rather greater weight is given to British data. While this book was being written, Norris and Vaizey's contribution[3] appeared. Their treatment is in some areas similar to mine, but it is rather less detailed. The coverage of the two books also differs.

As I started to write the book – a task somewhat more difficult than writing a synopsis – I realised that if it was to remain manageable, a number of topics would have to be excluded, or dealt with in summary fashion. This realisation came rather earlier to those who read and commented on the synopsis! Theorists will probably hold up their hands in horror at the failure even to mention some of the theoretical work on technological change that has been done at both the micro and macro levels. Management specialists will note the absence of detailed discussion on R & D (research and development) appraisal within the firm. Development economists will be surprised that the problems of low-income countries in relation to the availability of suitable technologies receive no mention.

In my view, however, these subjects are already dealt with in a fairly comprehensive way elsewhere, either in books or in survey articles. While their inclusion would probably have added to the completeness of the book, the end-product would have been unwieldy and rather less suitable for the undergraduate market and for the majority of non-specialists who might be attracted to the subject.

The inclusion of the case study on hovercraft development in part 3 does, of course, have an opportunity cost. But this development illustrates many of the issues and problems raised in the two previous parts so clearly (see below) that its place is more than justified. Indeed, a substantial portion of the whole book could have been written *around* this case study. The history of the hovercraft is also of interest in its own right as an important example of British inventiveness.

The plan of the book

Part 1 examines some of the more general issues associated with the economic analysis of technological change. Chapter 1 provides the background to the rest of the book. It looks at the expansion of R & D expenditures in the twentieth century and at the importance of technology in economic growth. This chapter also provides a brief review of the terms used in the book and of the economic characteristics of the main activities considered. Chapter 2 selects one of these activities, invention, for more detailed analysis and assesses how far it is an activity that can be explained by economic forces. The patent system is also discussed. Chapter 3 examines the relationship between the size of firm and R & D spending, invention and innovation. The existence of economies of scale in R & D is examined in the light of empirical evidence.

Chapter 4 discusses the role of innovation in competition and its place in traditional microeconomic theory. A distinction is drawn between static and dynamic efficiency, and 'restrictive practices' are analysed in terms of their effect on the innovation process. Empirical studies on the relationship of market structure to innovation are drawn on in this chapter. Technological barriers to entry are also examined.

The most important resource in the process of technological change is manpower. In chapter 5 the nature and operation of the market for qualified scientists and engineers (QSEs) is discussed, particularly in relation to shortages of different categories of personnel. The mobility of technical personnel between organisations is then examined. The analysis is broadened in the last part of the chapter to include an examination of the economic issues involved in the international migration of QSEs.

In part 2 the major channels for the production, dissemination and use of scientific and technological knowledge are discussed. In chapter 6 the government's financing and use of R & D resources, and the economic principles behind the public support of R & D are covered. Government policies and agencies designed to stimulate technological change are also analysed. Examples of the use of social cost benefit analysis in the field of R & D are provided. The role of industry is examined in chapter 7. Differences in research intensities and in the patenting activities of industries, and the interaction between innovation and industrial structure are also discussed. The last part of the chapter is devoted to the diffusion of innovation through industries. Chapter 8 examines the place of co-operative organisations in industrial R & D and the problems these bodies face. Chapter 9 concludes part 2 with an analysis of the role of the universities in scientific and technological change.

In part 3, as already mentioned, I report on my own research into the economic development of the hovercraft and associated craft and equipment in the UK. The exploitation of this invention has involved individual and corporate inventors, large, small and new companies, successful and unsuccessful companies, and government finance, through both government departments and the NRDC. As such, it should provide a highly useful complement to parts 1 and 2.

In chapter 10 a general background to the study is provided. Chapter 11 provides a fairly comprehensive picture of the scale of production in the industry and of the markets for hovercraft. The role of the individual inventor is discussed in chapter 12, while the technical development and exploitation of the air cushion principle is traced in chapter 13. Chapter 14 analyses the development of the industry on both the manufacturing and the operating sides. The involvement of the NRDC and government is discussed in chapter 15. Chapter 16 concludes the case study. The book ends with a few suggestions for further research work.

In applied work of this kind, it is almost inevitable that some sections will be a little out of date even before the book is published. New books, articles and statistics – both official and unofficial – on invention, innovation and R & D are constantly appearing; and in the hovercraft industry there is a continuous stream of new developments, of both a technical and a commercial nature. However, the only way of avoiding this problem is by not publishing at all – a solution that is even less satisfactory than the problem it is designed to solve.

Acknowledgements

I am grateful to several colleagues in the Department of Economics, Durham University, who either read or commented on parts of the manuscript or who helped me to unravel problems over which I had particular difficulty. I would especially like to thank Joost van Doorn, Mike Burns, Bernard Hall and Dick Morley for their assistance in this respect.

I am grateful to the Nuffield Foundation for providing the resources necessary to carry out the study on which part 3 is based, and to the University of Durham for providing some help in meeting typing costs. I am also very appreciative of the assistance given by the many individuals in industry and elsewhere who gave up their time to discuss the development of hovercraft with me. Many of them also commented on the first draft of part 3.

None of those who helped me is of course responsible for any of the errors or omissions found in the text.

Mrs M. Kimmitt and Mrs G. Gibson did an excellent job in typing the manuscript and I am most appreciative of their patience. I am also indebted to the Controller of Her Majesty's Stationery Office for permission to reproduce the statistics found in tables 3.3, 3.4, 6.2 and 7.2, and to a number of other publishers and authors who have kindly allowed me to use material in which they have the copyright. The sources of such material are fully acknowledged in the text.

Finally, I must acknowledge the debt I owe to my wife. Her encouragement and help were crucial to the successful completion of this book.

P. S. J.
The University of Durham
September 1974

REFERENCES

1. E. Mansfield *The Economics of Technological Change* New York, Norton (1968)
2. R. R. Nelson *et al. Technology, Economic Growth and Public Policy* Washington, Brookings Institution (1967)
3. K. Norris and J. Vaizey *The Economics of Research and Technology* London, Allen and Unwin (1973)

PART I

Some Economic Issues

1 Technological Change

The focus on the ways in which new or improved methods, organisations, products and processes are conceived, developed and harnessed for industrial use has become increasingly intensified in the twentieth century, and particularly in the postwar period. The first two sections of this chapter will discuss some of the causes of this growing emphasis and will provide a setting for the rest of the book. The third section looks very briefly at the attitudes of economists to technological change. The last section is devoted to a discussion of some of the more important terms used in this book.

1.1 The growth of research and development expenditure

Formally organised research and development (R & D), an important, though by no means the only, generator of technological change, has grown very rapidly in the last seventy years. Accurate estimates of this growth are very difficult to come by because it is only in recent years that statistics on R & D expenditures have been collected at all systematically. Even these statistics have shortcomings as measures of R & D *activity*. However, an impression of the kind of growth that R & D has experienced can be gained from the following estimates. In the UK, R & D expenditure (in 1963 terms) was probably less than £6m in 1900; in 1964–65 it had risen to around £750m (again in 1963 terms). This amounted to an increase of at least 125-fold.* In the United States, one estimate suggests that the real increase in R & D expenditure between 1930 and 1958 alone was twenty-five- to thirty-fold.[1]

---

* The 1900 estimate is based on a figure of 0.05 per cent of GNP spent on R & D in that year, given in D. W. Hill, *The Analyst* (1962).

This growth has been reflected in the activities of industry, government and the higher education sector. In industry the sectors that have experienced the most rapid output growth are the 'science-based' industries, eg electronics, chemicals, scientific instruments. Each of these now has a heavy commitment to organised R & D. (In 1900 few industrial research laboratories existed.) The 'traditional' industries, on the other hand, eg shipbuilding, cotton textiles, leather and clothing, have stagnated or declined and their R & D spending is relatively low (see p. 157). Even in the latter industries, however, many companies have accepted the need for R & D in the modern competitive battle. R & D, and its contribution to technological change, is now regarded as having major significance in the generation of new markets, and in the maintenance of competitive strength. It is often a large-scale consumer of a firm's resources. This in itself has generated new and special problems in industry: the Rolls-Royce collapse in 1970 in Britain showed only too clearly the effect of a company over-reaching itself in R & D, by the commitment of large financial resources to new and expensive developments which bring no immediate return.[2]

Government expenditure on R & D, encouraged by the massive appetite of military and space requirements, has risen from an almost zero level in 1900 to over fifty per cent of current expenditure on R & D in Britain and the United States (chapter 6). Whereas in 1900 there were virtually no government institutions with responsibility for supporting R & D, there is now a mass of government departments and laboratories that operate in this field. Major expansion of R & D commitment has also been seen in the universities. At the end of the nineteenth century the UK had only 287 university staff in the field of mathematics, science and engineering;[3] the universities now have well over 13,000 staff in these subjects.[4] These figures relate of course mainly to teaching staff, who also carry out research, although it is likely that the research element alone has increased at least at the same rate and probably higher.

One factor favouring an increase in R & D spending, especially in industry, has been the view that scientific and technological advances can be institutionalised, and that while such advances in the eighteenth century and during and immediately after the Industrial Revolution were usually made by men with little scientific training, and/or with very limited resources, modern developments of any importance are now a result of minds formally trained in the sciences, working often in large industrial laboratories. Several writers, including Schumpeter and Galbraith (p. 58) have held the view for example that it is no longer true that the majority of modern technical opportunities for investment come by chance at irregular intervals, but that they are now the planned consequence of considered decisions to finance R & D activity. In this way the speed at which advances emerge is seen as depending closely on the number of qualified R & D personnel working on them, and on the available technical facilities. This view is aptly summed up by Landes in his historical study of technology in the West : '. . . man can now order technological and scientific advance as one orders a commodity.'[5] There is clearly a good deal of truth in this proposition and it is supported by experience in the Second World War, where massive inputs of scientific resources often achieved the specified objectives. The development of the atomic bomb is perhaps the best example here.[6] However, it is by no means universally true : major 'accidental' discoveries are not unknown (see p. 215–6), and individuals working by themselves and small firms are still important sources of invention (see p. 59). Furthermore, it is the *relationship* between inputs and output – the productivity of

Table 1.1
Total expenditure on R & D (eleven countries)

Country	Year	R & D expenditure (US $m.)	R & D expenditure as % of National Income (market prices)
Belgium	1969	291	1.4
Canada*	1970	1092	1.6
France	1970	2650	2.0
Germany	1970	3550	2.1
Italy	1970	890	1.0
Japan	1970	3791	2.2
Netherlands	1970	696	2.4
Norway	1970	123	1.3
Sweden*	1970	368	1.5
UK*	1968	2424	2.6
USA†	1970	27,121	3.1

* Not including data for social sciences, law, humanities, education and arts.
† Not including data for humanities, education and arts.
Source : *UNESCO Statistical Yearbook 1972* New York, Unesco (1973) table 68
UN Statistical Yearbook 1972 New York, UN (1973) table 189

R & D expenditure – that is important, not merely that more input leads to more output. The productivity of R & D in different institutional contexts is examined in chapter 3.

Some idea of the commitment of the principal OECD member countries to R & D around 1970 is given in table 1.1 above. (The countries mentioned therein account for nearly all the R & D carried out in the non-communist world.) Conversion of currencies into US dollars at official exchange rates gives a somewhat distorted picture of the importance of R & D expenditure in different countries because these rates do not always reflect the true differences in the comparative costs of R & D. For example, Freeman and Young suggested in 1962 that a 'research exchange rate' of $6.30 should replace the (then) official rate of $2.80 when British–American comparisons of R & D expenditures were made.[7] This is a substantial difference; unfortunately the necessary refinement to table 1.1 cannot be made because of lack of data. Definitions of what is and what is not 'R & D' also vary between countries. These factors however are unlikely to affect the clear picture that emerges* of the heavy commitment of advanced countries to R & D and of the overwhelming dominance of the United States in the R & D effort of the non-communist world. In contrast no less developed country spends more than 0.5 per cent of its GNP on R & D.

1.2 R & D, economic growth and productivity change

Although the advanced countries tend to have relatively higher R & D expenditures compared with those of less developed countries, this does not of course necessarily imply a close link between such expenditures and *growth*. The absence of any close relationship is implied in the figures in table 1.1 : for example, both the UK and the USA, with relatively high R & D expenditures, have grown much more slowly than, say, Japan or West Germany.

The reasons for the absence of a direct link between R & D and growth are not hard to find. Firstly, 'R & D', as we shall see, covers

---

* The UN acknowledges the shortcomings of the data mentioned in the text, but takes the view that the figures do 'have some limited value in indicating at least a gross order of magnitude' (UNESCO *Statistical Yearbook 1972* New York (1973) p. 589).

a wide spectrum of activities which vary in their growth potential. Basic research, for example, will usually affect growth only in the long term if at all, whereas applied research and development may have much quicker results. Furthermore, basic research tends to favour all countries irrespective of whether or not they actually undertake it (see section 1.4 below). Again, R & D in some fields may not be as productive in growth terms as in others. The results of R & D directed towards final product, as opposed to process, innovations are unlikely to be fully reflected (if at all) in the measured growth figures (p. 13). Military and space R & D may have little growth potential, at least in the short run. Some research, eg into ageing, may reduce the possibilities for growth. The distribution of R & D between such activities varies considerably between countries.

Secondly, R & D *alone* is not sufficient to stimulate growth even though it may have the potential to do so. The results of such work must be translated into new products and processes. This requires numerous management skills in production, marketing, finance etc, which may be unrelated to those found within laboratories. Countries are likely to vary in their endowment of such skills. The relationship between R & D expenditure and growth therefore is likely to be a highly complex one, and to differ from country to country. At the *industry* level, the complications outlined above have less force and there does in fact appear to be a fairly close relationship between an industry's output growth and its R & D expenditure (p. 161). However, the causal direction is not clear.

While it may not be possible to identify the contribution of R & D to economic growth by a direct statistical test, a number of important empirical investigations into the growth process in advanced economies (often referred to as 'residual factor studies' for a reason that will become apparent later) in the postwar period have concluded that the increases in factor productivity as opposed to increases in the quantities of factors themselves have played a major part in economic growth. These productivity increases have been labelled by some economists as 'technical progress', although they clearly include the contributions made by many other factors apart from R & D.

Before looking briefly at the results of one of these studies, we ought first to examine the approach used since the results obtained can only be accepted if the approach is itself considered valid. The starting point for these studies has usually been an aggregate production function of the form :

$$Q = AN^\alpha K^{1-\alpha}$$

where Q, N, K are aggregate indices of output, labour and capital respectively. There are, of course, immense problems in aggregating heterogeneous units of both output and inputs. Capital is particularly difficult to measure in this way. This problem is usually tackled by weighting the different input and output units by their prices, although this solution in turn raises numerous other problems.* The production function above also assumes constant returns to scale (the exponents sum to one). Thus a one per cent increase in all inputs leads to a one per cent increase in output. The individual exponents provide a measure of the elasticity of output with respect to the relevant factor. Thus, if N increases by one per cent then output will increase by (approximately) α per cent. Under perfect competition, these exponents are in fact the relative factor shares. This may be shown for example in relation to labour. The marginal product of this factor is :

$$\frac{\partial Q}{\partial N} = \alpha A N^{\alpha-1} K^{1-\alpha}$$

and the total payment it receives is therefore $N \frac{\partial Q}{\partial N}$ (under perfect competition, labour is paid the value of its marginal product). Since,

$$N \frac{\partial Q}{\partial N} = \alpha A N^\alpha K^{1-\alpha}$$

$$= \alpha Q$$

the share of labour in total output must be

$$N \frac{\partial Q}{\partial N} \Big/ Q = \alpha \quad .$$

The approach adopted in the residual factor studies is to calculate the *actual* growth rate of an economy over a given period; then to estimate what growth of output would have been if the growth of labour and capital over the period were weighted by their relative

* For an analysis of some of these problems, see C. Kennedy and A. T. Thirlwall, reference cited in n.8.

factor shares. The differences between these actual and hypothetical growth rates is the residual denoted by r below.

$$r = q - \alpha n - (1 - \alpha) k$$

where q, n and k are the rates of growth of output, labour and capital respectively. Alternatively, the residual may be interpreted in terms of that part of the growth in output per head that cannot be attributed to increases in capital per head, weighted by the share of capital in output, that is :

$$r = q - n - (1 - \alpha)(k - n).$$

As noted earlier, the studies using basically this approach have pointed to a very substantial residual. Work by Solow and others for example suggests that between eighty and ninety per cent of growth in output per head in the United States in modern times could not be accounted for by growth in capital per head.[8] Other studies in other countries also point to very large residuals.[9] The basic question is, how should this residual be interpreted? Strictly speaking, as Abramovitz has pointed out, 'it may be taken as some measure of our ignorance about the causes of economic growth . . .'.[10] It has however been commonly attributed to 'technical progress', which is a catch-all phrase for anything that might conceivably alter total factor productivity. Griliches and Jorgenson have however argued [11] that the large size of the residual primarily reflects errors in measuring the factor inputs. Clearly, how these are measured will directly affect the size of the residual. In their study for example they showed that in the US private domestic economy growth of total productivity 'explains' only 3.3 per cent of output growth between 1945 and 1965 compared with only 47.6 per cent before 'correcting' the data for quality changes in factor inputs, etc. But their study in turn raises a number of questions of measurement and methodology. As Kennedy and Thirlwall point out, the approach adopted in growth studies must depend on the purpose of the study. For example, 'If the purpose is simply to measure the increase in the productivity of factors of production over time, it makes no sense to adjust the factors for quality changes.'[12]

Perhaps the most far-reaching residual factor study to date is that by Denison,[13] who not only adjusted the factor inputs for certain changes, eg in age–sex composition of labour, hours of work – this in turn of course changes the size of the residual – but

Table 1.2
Percentage distribution of growth rates of adjusted National Income[a] among the sources of growth, 1950–62, for the US and NW Europe

Source of growth	US %	Belgium %	Denmark %	France %	Germany %	Netherlands %	Norway %	UK %	Total NW Europe %
Adjusted National Income[a]	100	100	100	100	100	100	100	100	100
Total factor input	58	39	46	26	38	42	30	47	36
of which: Labour	33	25	18	10	19	19	4	25	18
Capital	25	14	29	17	19	23	26	21	18
Output per unit of input	42	61	54	74	62	58	70	53	64
of which: Improved allocation of resources and economies of scale	20	34	39	41	40[b]	31	43	20	36[c]
The 'residual' (ie advances of knowledge and changes in the lag in application of knowledge general efficiency, and errors and omissions)[d]	23	28	15	32	22	27	27	33	28

[a] The adjusted figures are arrived at by deducting from the actual growth rates the estimated effect of (i) irregular fluctuations in farm output and (ii) incomparabilities between terminal years with respect to the intensity of resource utilisation. These deductions are made by Denison to improve comparability of growth rates between countries.
[b] Includes 4 percentage points for balancing of capital stock.
[c] Includes 2 percentage points for balancing capital stock.
[d] As mentioned in the text, Denison assumes that the residual in the US countries consists solely of advances in knowledge.

Source: E. F. Denison, assisted by Jean Pierre Poullier *Why Growth Rates Differ, Postwar Experience in Nine Western Countries* Washington, Brookings Institution (1967) chapter 21. Detailed figures may not add to subtotals because of rounding.

also attempted on the basis of certain fairly strong assumptions to break down the growth of total factor productivity into a number of component parts. His findings for the 1950–62 period for Europe and the US are summarised in table 1.2. It is the last row of figures that is of particular interest to us here. Denison does in fact break this figure down further by assuming that the US residual (0.76 percentage points or 23 per cent of the US growth rate, 1950–62) is wholly attributable to advances in knowledge. He makes no allowance for changes in the lag in the application of knowledge or for errors or omissions for this country. He then argues that, as knowledge is an international commodity, the same contribution in percentage points would be made by advances in knowledge to growth rates in countries in north-west Europe. What remains of these countries' growth rates after this deduction of advances in knowledge is then allocated to 'changes in the lag in the application of knowledge, general efficiency and errors and omissions'. We have grouped these two elements (ie advances in knowledge and changes in the lag, etc) together in the table because the assumptions on which their separation is made seem to be particularly strong. In any case, in the majority of countries, according to Denison, the second factor is very small indeed.

Two features stand out from the data given in the table. Firstly, increases in factor productivity from whatever source accounted for over 50 per cent of total growth in all the European countries mentioned. This figure, it must be remembered, is arrived at *after* adjusting factor inputs for certain quality changes, adjustments that have not normally been carried out in other studies. It is also of course dependent on the assumption that the individual exponents in the production function can be equated with the relative factor shares (p. 81). Secondly, in every country except Denmark, the 'core' residual, ie the last row in the table, accounted for over one-fifth of total growth, a very considerable contribution.

These results must be interpreted in the light of the basic assumptions made by Denison on both the theoretical and the empirical level. Some of these assumptions are of a highly arbitrary nature. For example, Denison's estimates of the contribution of economies of scale to the growth of total factor productivities are little more than informed guesses. As he says of his estimate for the US economy (on which much of his analysis for Europe relies):
'My assumption was merely an attempt to set down a figure that

was at the same time what might be termed a qualitative weighted average of views expressed by other economists and a figure that seemed reasonable to me.'[14] Some economists argue that Denison's work is open to too many criticisms for his estimate of the residual to be accepted.[15] It must also be remembered again that much R & D effort – eg that directed towards new and improved products, which change the quality of output – may not be reflected in the measured growth figures. Indeed, Denison himself argued in an earlier analysis of US growth that only about one-fifth of the contribution made to growth by 'advances in knowledge' could probably be attributed to R & D expenditures.[16] Nevertheless, his work does provide concrete estimates of particular components of growth and it does suggest even if the margin of error is wide, that 'technical progress' at which some R & D is aimed is likely to be an important source of measured growth.

At the industry level, Matthews has undertaken a residual factor study for Britain. His results are shown in table 1.3. His residual is approximately equivalent to the 'output per unit of input' source of

Table 1.3
The residual in output growth in manufacturing industries in Britain, 1948–68

Industry	Output Growth (1)	Residual (2)	(2) as % of (1)
Chemicals	5.68	2.42	42.6
Metals	1.66	0.01	0.6
Engineering and shipbuilding	3.88	2.59	66.7
Vehicles	4.55	3.15	69.2
Textiles	1.31	3.04	195.8
Leather, fur, clothing	1.27	1.55	122.0
Bricks, etc	3.29	1.39	42.2
Timber	2.45	1.90	77.5
Paper, etc	4.41	2.24	50.7
Other manufacturing	4.32	2.05	47.4

Note: Two industries, Food Drink and Tobacco and Metals n.e.s., had *negative* residuals of 0.35 and 0.78 respectively.

Source: Adapted from R. C. O. Matthews 'Some Aspects of Postwar Growth: the British Economy in Relation to Historical Experience' *Manchester Statistical Society* (1964), extended by A. J. Buxton. Quoted in A. R. Prest *et al., A Manual of Applied Economics* London, Weidenfeld and Nicolson (1972) p. 158

growth in table 1.2 (Buxton has shown that if factor inputs are adjusted for quality changes, etc, along the lines suggested by Griliches and Jorgenson, the residual does not always fall.[17]) Again the residual is, with few exceptions, very important, in two industries more than offsetting the effects of a decline in factor inputs. What is particularly interesting is that in several declining industries the contribution made by the residual has been considerable. Textiles, leather, fur and clothing are particularly noteworthy in this respect. This may suggest that industries that are not growing or are stagnant have an incentive to improve their technology at least to maintain their position.

1.2.1 R & D and productivity in industry

Several investigators have looked directly at the relationship between R & D and total factor productivity in industry. Although it seems reasonable to suppose that R & D is likely to affect productivity, this relationship may be difficult to identify empirically for several reasons. (There are also a number of technical problems relating to measuring the variables involved. These issues are however ignored here.) Firstly, productivity is affected by a whole host of other factors apart from R & D. Growth of output itself is likely to lead to increases in productivity through scale effects. Secondly, R & D expenditures are likely to affect productivity only after a lag. This lag will vary across projects and industries. Thirdly, as we have already noted, some industrial R & D is devoted to new products rather than processes and may not therefore be reflected in the productivity measures. R & D expenditure on consumer goods may not show up in the latter at all. In the case of new products used in the production processes of another industry, the productivity of the *buying* industry will increase, whereas the R & D will have been carried out in the *selling* industry. To attempt to relate R & D and productivity in the same industry in this case would therefore be misleading. Fourthly, industries may vary in the ease with which a given productivity increase can be achieved. This variation may occur over time as well as across industries.

The empirical work undertaken in this field has generated mixed results, as would be expected from the above discussion. Minasian has made a study of the relationship between R & D expenditure and total factor productivity increase for a sample of eighteen firms in the US chemicals industry over the period 1947–56.[18] He regressed the rate of growth of productivity on average R & D expenditures deflated by a weighted average of total inputs, or by average gross plant and equipment, lagging R & D by up to five years. His best results were obtained when productivity over the period 1947–57 was regressed on average R & D expenditure deflated by average gross plant and equipment for the same period: the r^2 was 0.73. Minasian also found that, almost without exception, the shorter the period over which the change in productivity is measured, the lower the r^2. He argues that, as the period considered shortens, so the influence of 'abnormal' factors becomes relatively greater. Minasian's results for the longer period suggest that R & D exercises an important influence on productivity growth. He is careful to point out however that because of the nature of his sample his results cannot be regarded as having general applicability.

Taylor and Silberston have conducted a more general exercise into the relationship between total factor productivity and R & D intensity for twenty-six UK manufacturing industries.[19] They found an r^2 of only 0.37 when the increase in productivity between 1958–63 and 1963–68 (using the average figure for each period) was regressed on the average R & D/net output ratios for 1958–63. They also found that net output growth was more strongly correlated with productivity growth than with R & D intensity. It seems from their analysis that much of the apparent effect of R & D intensity on productivity comes in fact from output growth with which R & D intensity is also correlated. Output growth may of course enable the results of R & D to be applied (via new investment opportunities), thereby raising productivity. However, it is clear that, although there may be some support for R & D leading directly to higher productivity, the exact relationship involved is complex. Productivity is clearly affected by numerous other factors.

Despite our limited knowledge on the exact contribution made by R & D to economic growth and the increase of productivity, there is undoubtedly a widespread conviction that it is large. The emphasis on growth as an objective of public policy has thus led to the advance of science and technology and ways of organising and stimulating it in both the public sector and private industry becoming an important focal point for discussions on economic policy. The assumed technological 'gap' between the US and Europe, particularly Britain, has received widespread attention.[20]

And in the 1960s, 'a white-hot technical revolution' was to spearhead a revival of Britain's growth record.[21]

1.3 The economist and technological change

The attention given to the process of technological change in economic thought has varied considerably. At first sight, the mainstream literature of classical economics appears to give little prominence to technological knowledge and its application. Robbins however has challenged this view, arguing that, although at first such a conclusion appears justified, a more detailed examination of the writings of the classicists shows that they were certainly aware of the importance of the advancement and utilisation of knowledge in economic development.[22] Whatever the classicists' approach, there can be little doubt that the neo-classicists gave scant attention to technological change. The microeconomics developed by them, which still forms the basis of much of university teaching in this area, incorporated the assumption of technology being largely taken as given, any change being treated as exogenous to the system. It is hardly surprising therefore that this area of economic theory has become the happy hunting ground for modern economic iconoclasts such as J. K. Galbraith.*

In another main development of economic theory in the twentieth century – Keynesian analysis – technology is also assumed to remain constant. Keynes's own formulation of the assumptions on which his theory was based indicates this clearly: 'We take as given . . . the existing quality and quantity of available equipment, the existing technique . . .'.[23] Because of the time period Keynes was considering, this assumption was not unrealistic. Unfortunately, however, some postwar growth models, which must inevitably be concerned with much longer periods, have attempted to 'dynamise' Keynes, without recognising the essentially short-run nature of his assumptions.

---

* See, for example, his *New Industrial State*, London, Hamish Hamilton (1966), p. 46 fn: 'Changing technology, it is conceded, alters progressively and radically what can be obtained from any given supply of factors. But there is no way by which this intelligence can be developed at length in a textbook. So economic instruction concedes the important, and then discusses the unimportant.'

In recent years technological change has received greater prominence in growth and development theories, and the empirical studies by Denison and others mentioned earlier have given added emphasis to the subject. Yet the *institutional* side of R & D, invention and innovation, and the economic influences on their supply, demand and distribution, have been examined in what Salter called in 1958 a 'highly sketchy fashion'.[24] Apart from the work of Carter and Williams, Jewkes, Freeman and a few others, the empirical cupboard in the UK at least is still relatively bare. The picture is gradually changing, but there is still much to be done, on both the theoretical and empirical front. As Schmookler says, technological change is still the *terra incognita* of modern economics : 'Economists have only the most general ideas about what determines it . . . and some of those ideas are wrong'.[25]

One of the reasons why economists have paid relatively little attention to scientific and technological change and the activities that lead to it is probably due to the high degree of uncertainty associated with such activities; economists find things much easier when operating in a world of certainty! Furthermore, much of the subject matter may be incomprehensible to the economist. This second point raises a whole range of fundamental issues relating to the role of scientists and technologists in decision-taking at the firm and national level.[26] The inability of non-scientists to comprehend detailed scientific issues – which may in some cases constitute the essence of a decision – places the person who is 'in the know' in a very much stronger position. Nevertheless, despite these difficulties the vast sums involved in R & D (and the clear mistakes that have been made in their allocation) make it imperative that the economist bring his skills (either in their existing or in an adapted form) to bear in this field. This book attempts to lay a foundation for this by drawing together some of the theoretical and empirical work that has been done so far. Before this is attempted in the succeeding chapters, however, the definitions of the most important terms used in the book are outlined below.

1.4 Some definitions

1.4.1 *Technology*

Technology has been defined by Mansfield as 'society's pool of knowledge regarding the industrial arts'.[27] This is a useful working definition, although it does raise problems in the use of certain tools of analysis in economics, such as the production function. This function specifies the relationship between factor inputs and the (maximum) output obtainable from any combination of such inputs, *given the available technology*. But what does this last phrase mean? If, as Downie points out, a library of blueprints was freely available to all showing 'the most appropriate technique for all conceivable production objectives',[28] then there would be little difficulty. In practice, however, the given state of technology is much harder to define. Some processes are embodied in the machines and equipment already on the market. But there may also be a whole range of possible techniques that have not reached that stage but are clearly feasible in the technical sense. They may even be on the drawing board. Rarely however will these techniques be embodied in machinery and equipment *without* modification, although the need for the latter may only become clear *after* the construction of the machinery has been started. Another complication stems from the fact that some techniques are known only to certain firms, either because of industrial secrecy or simply because information is unevenly distributed. Furthermore, as Eckaus has pointed out, decision-takers may *think* they are aware of all the production possibilities, when in fact they are not.[29]

Schmookler has drawn a distinction between production technology and product technology.[30] The former relates to the knowledge used to produce a good that is embodied in the materials, machines and processes used in fabrication. Product technology on the other hand defines the knowledge used in creating and improving the products themselves. Hence a change in production technology might leave the final product virtually unchanged, while a change in product technology could be achieved without any change in process technology (ie, if a change in the final product can be accomplished without any modification of the firm's capital equipment). In practice however these technologies are closely intertwined. Changes in product technology are usually heavily dependent on changes in the relevant processes: few product alterations can be made without a change in techniques of production. Furthermore, one industry's product may be another's raw material or be closely connected with its processes. Sheet steel, for example, one of the iron and steel industry's end-products, forms the basic material for many other industries, eg motor, aircraft, etc. Machinery represents the end-product of the engineering industries, yet it is used in the processes of most other industries. It is this kind of overlap which in many cases leads to the production technologies of one industry consisting, to a large extent, of the product technologies of others.

1.4.2 *Science*

Although at the margin the distinction between science and technology is a blurred one, in general there is a clear difference in the areas these terms seek to define. Briefly, science may be said to be the sum total of systematic and formulated knowledge about the natural world, whereas technology is use-orientated.*

1.4.3 *Invention*

A major component of the process of technological change is invention, the creation of an idea and its first reduction to practice in a physical form. To Schumpeter an invention produces 'of itself no economically relevant effect at all'.[31] Its newness is of course an essential characteristic, although it will of course utilise existing knowledge, perhaps in a new form. Important inventions are usually associated with a mass of 'minor' ones, which enable the former to be used commercially. For example, although the first spindle-type cotton picker was first produced in 1889 in the United States, it was not until 1948 that it was introduced on a commercial basis. Part

---

* This should not be taken necessarily to imply that technology is simply the derivative of science, ie that it is 'applied science'. Price for example has argued that 'old technology breeds new', just as scientific advances build on previous science. See his 'The Structures of Publication in Science and Technology' in W. H. Gruber and D. G. Marquis (eds) *Factors in the Transfer of Technology,* Cambridge, Mass.: MIT Press (1969).

of the reason for this delay was economic, but much of the time was spent by the inventor and firm that took it up developing the invention technically to the stage where it became a commercial proposition.[32]

Schmookler has suggested that a definition of invention might include prospective utility as a necessary component.[33] This qualification however would nonetheless create numerous definitional problems, as well as invalidating many thousands of legal inventions that have never been used. An invention should be distinguished from a scientific discovery. The latter aims at revealing what *is* already, while the former involves a human agency in the creation of a physical form that did not previously exist. This distinction comes out clearly in the development of the fluorescent lamp. By the mid-nineteenth century it had been shown that ultra-violet radiation could produce fluorescence in certain materials and that an electric discharge through mercury under low pressure could produce ultra-violet radiation. It was on the basis of these scientific discoveries that inventive activity to produce a suitable lamp was pursued.[34]

1.4.4 *Innovation*

An invention may not be taken up industrially for a number of reasons : it may not be profitable; though profitable, the firm may have wrongly estimated the likely returns; or the firm may be unable to take it up for a variety of reasons, such as a shortage of capital or limited management capability. However when an invention, either a new one or one that is old but has not previously been taken up, is utilised by industry and exploited commercially, 'innovation', in Schumpeter's terminology, occurs. To Schumpeter, who has probably done more than any other economist to focus attention on the innovation process, invention and innovation were conceptually quite distinct. This distinction is usually maintained in the literature, although it has been criticised on the grounds of artificiality : that it splits what is essentially a unified process of bringing new things into production (see p. 30). And as shown above, the very act of innovation often requires further invention to enable the original invention to be utilised commercially. Here, in this process, invention and innovation become inseparably interwoven. Schumpeter's distinction between the two may only be really useful therefore when basic or *primary* inventions are being considered.

Schumpeter's discussion of innovation is couched in terms of *commercial* exploitation. However, there are numerous changes in end-products and methods of operating in non-commercial fields, eg health and defence, that might also be included under the term 'innovation', and be similarly subjected to economic analysis. Much of the discussion in this book however will be devoted to innovation in the context of private enterprise. Schumpeter did however see the innovation process as embracing much more than the introduction of new processes and products and as including new markets and new forms of organisation – 'in short, any "doing things differently" in the realm of economic life'.[35] The discussion in this book will again be limited and will not generally extend to the last two types of innovation, although new products and processes will often imply new markets and forms of organisation.

Innovation here is limited to the *first* firm to introduce a product or process change. Other firms who copy or follow this innovating firm may be regarded as *imitators*. This distinction is useful from an economic viewpoint, since not only does it correspond with the distinction between the growth and the dissemination of innovations, but it enables two activities that involve somewhat different stimuli and pressures to be treated separately.

Innovation can be extremely risky. Carter and Williams for example have shown that there are considerable differences between the *expected* and *actual* returns on investment projects; in fact, for their sample of firms in UK industry, the correlation between the two was only 0.13.[36] Demand and cost changes played a substantial part in upsetting the estimates.

1.4.5 *Research and development*

R & D, as shown earlier, has become an important user of resources. It covers a wide range of activities that are very differently motivated. Because of the lack of distinct separation between these activities, reference will be made to the R & D 'spectrum'.

Not surprisingly, it is difficult to obtain complete agreement between writers on exact conceptual definitions for the various activities involved, but at the risk of over-simplification the spectrum is divided broadly into three.[37] Firstly, 'pure' research describes the examination of fundamental problems where no application is in view at the time of investigation. It is carried out primarily to increase scientific knowledge, there being no specific intention that the knowledge so acquired will become useful. 'Applied' research relates to investigations that have a definite application in mind at a foreseeable time. Between these two may be placed a twilight category of 'basic' research, which, while not undertaken with a specific application in mind, is nevertheless done 'in fields of recognised technological importance'.[38] In this book, 'pure' and 'basic' research are grouped together under the latter heading. The difference between 'basic' and 'applied' research hinges essentially on the motivation of the research worker. (The distinction corresponds broadly with that between 'curiosity-oriented' (or non-mission) research and mission-oriented research (see p. 215). This distinction is used fairly widely in the United States.) As such it is inevitably a matter of degree. Few scientists will undertake research in an area that has no prospect of yielding results that will eventually have some 'use', however indirect or however long-term. The motivation of the financer of such work may however be different from that of the researcher. Several large companies for example finance a small basic research effort in particular areas in the hopes that it will eventually lead to commercially applicable results (see p. 217). 'Development' moves a stage further than applied research, and is largely devoted to bringing an invention to the stage where it can be produced commercially. If this phase, which may include pilot plant studies, is passed successfully, then productive capacity may be built and the product launched on to the market. Thus the spectrum ranges from the extension of knowledge for its own sake to the introduction, on to the market, of new or improved products.

The complexity of this spectrum must make any definition of its component parts very difficult, and any division will be lacking in some respect as a result. Each 'band' of the spectrum has a momentum of its own, and all three are often interlinked: applied research or development may require further basic work to be conducted in parallel;[39] it may also be difficult in practice to draw any definite line between basic and applied research and development. The researcher may himself also move along the spectrum in the course of his work.[40] But although the distinctions may be somewhat fuzzy, it will still be true that some projects are more (or less) basic than others.

The spectrum so described does not imply that if an individual (or a firm or nation) wishes to undertake applied research and/or development, he must first carry out the necessary basic research. It may be economically sound to carry out applied research and development obtaining the foundation for such, ie basic research, from sources external to himself[41] Even the development stage may be avoided – either by imitation of the products of other firms, or by the acquisition of production licences. For many firms, the unit costs of undertaking their own R & D may be higher than the alternative unit licence costs because of the limited markets they face. The risk may also be less: the results of a company's own R & D are likely to be rather uncertain, whereas a licence will often provide access to a proven technology. One country in which such licence agreements are very important is Australia: this is not unexpected, bearing in mind the fairly limited domestic market in this country and the relative youth of many of the industries established there.[42] It must be remembered however that even licence agreements need a certain element of expertise on the part of the licensee, especially if adaptions have to be made, and that some licences may simply not be available. Furthermore, a firm that produces solely under licence, without any R & D effort of its own, may find expansion difficult, particularly in overseas markets (p. 173).

Because the R & D spectrum figures so prominently in most of the following chapters, its basic economic characteristics are best dealt with at the beginning of the book. These are outlined below.

Basic research

Firstly, both the technical and commercial outcome of applying effort in basic research will usually be highly unpredictable, whereas the results of applied research and development may in some cases be comparatively fairly well established. Forecasts – if any take place – of future returns from work in the last two fields can be wildly wrong, especially where an entirely new product or process is involved; if investment predictions are usually very inaccurate (see above), then the evaluation of activities that *precede* such investment is hardly likely to be better. However, at least some assessment can usually be undertaken. Expenditure on basic research, on the other hand, is almost an act of faith. It is often impossible even to envisage commercial results – especially in 'pure' research : the mathematical inquiries into molecular physics, for instance, took place long before and without any foresight of the valve that led to broadcasting and long-distance telephone communications. Similarly, Faraday's explanation of electromagnetism on which the modern electrical industry is based was undertaken without a view to its future possible application.

Secondly, the period in which an 'investment' in basic research will pay for itself is usually so long-term that it will normally be outside the maximum period that most companies allow for an investment to become profitable. In 1958 a McGraw-Hill survey was conducted, which included almost every large company in industry and commerce in America, on the length of time in which these companies expected their R & D to pay off. Thirty-nine per cent replied that they expected their R & D to pay off in less than three years; another 52 per cent indicated an expected pay-off in three to five years; and 9 per cent put the expected pay-off at six years or more.[43] Thus, 91 per cent expected a pay-off from their R & D spending in no more than five years. Although no survey has been done on the length of time in which basic research pays off, in most cases it will normally be much longer than five years (see p. 218).

Nevertheless, the eventual total *social* benefits derived from basic research, if calculation was possible, are likely to be much greater than those from similar amounts of effort in applied research or development, since 'successful' basic research may stimulate several applied research and development projects. On these grounds, Nelson concluded that a firm is usually unable to exploit the full economic advantage of a basic research project because its findings will have a far wider application than is open to the individual firm to exploit.[44] (The same will be true on a national level.) The extent to which a firm can use basic research findings may depend in part on the width of the firm's technological base. A firm such as ICI or Du Pont, which embraces a field almost as wide as a discipline itself, is likely to have a far wider 'catchment area' for basic research findings than the small firm concentrating on a particular branch of chemistry. In the latter case, only a small percentage of the benefits derived from such work are likely to be expressed in the company's output. Research in such companies will probably be of a far more applied nature, although there is no reason why such a firm should not expand (provided it has or can obtain sufficient funds and expertise) in order to exploit its research findings.

At the same time, whatever the extent of a firm's interests, the private returns obtained from basic research are likely to diverge widely from total social returns. However, it should be pointed out that prospective private returns might still be regarded as *high enough* to induce expenditure of research of a basic nature by an individual firm. To illustrate this, Griliches gives the example of Du Pont which undertook the basic research necessary for the production of nylon, even though the total potential social returns were higher than the private returns, because the latter were *sufficient* to attract the investment (see also p. 217).[45]

Thirdly, the firm working in basic research is unlikely to be able to keep its research findings to itself although it may gain some advantages in being first in the field. The lead so obtained will be particularly important where an industry is heavily dependent on advances in basic research, eg in chemical and pharmaceutical products. Natural laws and fundamental principles are not patentable, and are therefore, (after a time lag), fed into society's common pool of knowledge. Restrictions on the publication of findings in basic research would evoke vehement opposition from the scientific professions, and any firm that adopted such a policy might be unpopular with its own research staff. This common pool is then available to be applied by any firm that is willing and able to interpret such findings, and in this sense, basic research work may be classed as a 'collective' or 'public' good – in that when one firm provides such knowledge for itself it automatically and unavoidably provides for others.

If basic research by its nature provides material on which a firm may base its applied research and development without incurring the initial costs then the incentive for any *one* firm to perform such activity is likely to be low, especially as the prospective returns are both distant and uncertain. And it is because private opportunities are unlikely to draw as many resources into basic research as is considered socially desirable that there will almost certainly be a case for some special financial arrangement to support it (chapter 6).

The 'free' nature of basic research findings also has important implications at the international level. A country may finance only a very limited basic research effort itself, yet be able to obtain the results of such research from other countries without charge. This may mean, as H. G. Johnson points out, that leadership in technology is not dependent on leadership in basic science.[46] Indeed, there is some very limited evidence to suggest that they may even be in conflict,[47] one reason for this being that basic research effort employs manpower and equipment that might otherwise be used for applied research and development. One qualification must however be added to the above argument: some basic research may have to be undertaken to provide a training and attraction for manpower that will spend most of its time on applied research and development. It may for example be argued with some force that only when a scientist has himself engaged in basic research can he appreciate the significance of the results obtained by others.

Applied research and development
At the other end of the spectrum, effort in applied research and even more so in development is directed towards the introduction, on to the market, of new or improved products and processes. It is therefore in the work conducted in the applied research and development laboratories that much of a firm's competitive strength and profitability will lie. Unlike findings in basic research, a firm can, within the limits of patent laws and commercial secrecy, ensure that the results of its laboratory work are not taken up by competitors (see chapter 2). Mueller's attempt to correlate the number of patents with expenditure on the component parts of R & D in various industries,[48] although resulting in some surprising conclusions, at least confirms that it is with expenditure in applied research and development, rather than in basic research, that the closest correlation usually exists. Partly because of this protection, and partly because of the very nature of applied research and development, the divergence between private and social returns is likely to be much less than in a basic research project.

Moreover, the commercial and technical outcome of applied research and development is usually more certain than that of basic research. Although a firm may be unable to forecast accurately the market response to an innovation, it has a number of market research methods available to it that can reduce the area of risk. Research directors are also likely to have some idea of whether particular technical problems or 'teething troubles' will be solved, and whether a specific research objective will be achieved. The degree of uncertainty in applied research and development will of course depend on the stage which the work has reached and on the nature of the product or process being developed.

It can be seen that the fundamental activities considered in this book – invention, innovation and R & D – to some extent overlap, making precise definition difficult. Nevertheless, despite difficulties at the margin, some distinction between the terms must be maintained since they are often subject to different economic stimuli and constraints. In the following chapter we turn to a more detailed discussion of one of these activities, invention.

NOTES on Chapter 1

1. L. S. Silk *The Research Revolution* New York, McGraw Hill (1962) p. 160
2. *The Times* (5 February 1971)
3. M. Sanderson *The Universities and British Industry, 1850–1970* London, Routledge & Kegan Paul (1972) p. 23
4. *Statistics of Science and Technology 1970* London, HMSO (1971) table 34
5. D. S. Landes *The Unbound Prometheus* Cambridge, Cambridge UP (1969) p. 538
6. L. S. Silk op. cit. p. 55
7. C. Freeman 'Research and Development: A Comparison between British and American Industry' *National Institute Economic Review* (1962)
8. See for example R. M. Solow 'Technical Change and the Aggregate Production Function' *Review of Economics and Statistics* (1957), and S. Fabricant 'Economic Progress and Economic Change' 34th Annual Report of the National Bureau of Economic Research 1954, quoted in C. Kennedy and A. P. Thirlwall 'Technical Progress' *Economic Journal* (1972)
9. Quoted in B. R. Williams *Investment, Technology and Growth* London, Chapman Hall (1967) chapter IX
10. M. Abramovitz 'Resource and Output Trends in the US since 1870' *American Economic Association (Papers and Proceedings)* (1956)
11. Z. Griliches and D. W. Jorgenson 'Sources of Measured Productivity Change' *American Economic Association (Papers and Proceedings)* (1966)
12. C. Kennedy and A. P. Thirlwall op. cit.
13. E. F. Denison *Why Growth Rates Differ* Washington, Brookings Institution (1967)
14. ibid. p. 226
15. See for example, J. Vaizey and K. Norris *The Economics of Research*

and Technology London, Allen and Unwin (1937) p. 100
16. E. F. Denison *Sources of Economic Growth in the United States* New York Committee for Economic Development (1962) p. 245
17. A. J. Buxton *An Examination of the Residual Factor in UK Manufacturing* University of Warwick, Centre for Industrial Economic and Business Research, Occasional Paper No 19 (1972)
18. J. R. Minasian 'The Economics of Research and Development' in *The Rate and Direction of Inventive Activity* Princeton, Princeton UP (1962)
19. C. T. Taylor and Z. A. Silberston *The Economic Impact of Patents* Cambridge, Cambridge UP (1973) pp. 66–78
20. See for example M. J. Peck 'Science and Technology' in R. E. Caves *et al., Britain's Economic Prospects* London, Allen & Unwin (1968); B. R. Williams *Investment, Technology and Growth* London, Chapman Hall (1967), chapter 1; D. Swann and D. L. McLachlan *Concentration or Competition: A European Dilemma?* London, Chatham House and PEP (1967); C. Layton, *European Advanced Technology* London, Allen & Unwin (1969). See also the author's 'Research in Britain Today' *Lloyds Bank Review* (1969)
21. See M. J. Peck in R. E. Caves *et al.,* op. cit. p. 448
22. L. Robbins *The Theory of Economic Development* London, Macmillan (1968), Lecture 4
23. J. M. Keynes *The General Theory* London, Macmillan (1936) p. 245
24. W. E. G. Salter *Productivity and Technical Change* Cambridge, Cambridge UP (1960), p. 8
25. J. Schmookler *Invention and Economic Growth* Cambridge, Mass., Harvard UP (1966) p. 3
26. D. K. Price *The Scientific Estate* Cambridge, Mass., Harvard UP (1965), and Gilpin *American Scientists and Nuclear Weapons Policy* Princeton, Princeton UP (1962)
27. E. Mansfield *The Economics of Technological Change* New York, Norton (1968) p. 10
28. J. Downie *The Competitive Process* London, Duckworth (1958) p. 84
29. R. S. Eckaus 'The Factor Proportions Problem in Underdeveloped Areas' *American Economic Review* (1955)
30. J. Schmookler op. cit. p. 88
31. J. Schumpeter *Business Cycles* New York, McGraw Hill (1939) vol. 1, p. 84
32. J. Jewkes *et al. The Sources of Invention* London, Macmillan (1969) p. 244
33. J. Schmookler op. cit. p. 6
34. J. Jewkes *et al.* pp. 254–6
35. J. Schumpeter op. cit. p. 84. See also pp. 87–8
36. C. F. Carter and B. R. Williams, *Investment in Innovation* Oxford, Oxford UP (1959) p. 90
37. The following description of the spectrum follows fairly closely the definitions adopted by the *Zuckerman Committee on the Management and Control of Research and Development* London, HMSO (1961)
38. Zuckerman op. cit. p. 7
39. See R. R. Nelson 'The Link between Science and Invention: The Case of the Transistor' in *The Rate and Direction of Inventive Activity,* Princeton, Princeton UP (1962)
40. See W. R. Maclaurin *Invention and Innovation in the Radio Industry* New York, Macmillan (1949), in which he quotes Oliver Lodge as an

example of a scientist in the radio industry who was interested in practical application of his work (p. xix). He also quotes other scientists, eg Faraday, Maxwell and Hertz, who did not concern themselves with the commercial fruits of their discoveries. Examples can be multiplied on both sides.

41. R. R. Nelson 'The Simple Economics of Scientific Basic Research' *Journal of Political Economy* (1959). See, however, F. M. Scherer 'Firm Size and Patented Inventions,' *American Economic Review* (1965)
42. P. Stubbs *Innovation and Research* Melbourne, Institute of Applied Economic Research (1968)
43. D. Hamberg 'Invention in the Industrial Research Laboratory' *Journal of Political Economy* (1963)
44. R. R. Nelson op. cit.
45. Z. Griliches 'Research Costs and Social Returns' *Journal of Political Economy* (1958)
46. H. G. Johnson 'Federal Support for Basic Research: Some Economic Issues' in *Basic Research and National Goals* Washington, National Academy of Sciences (1965)
47. See T. S. Kuhn's 'Comment' on I. H. Siegel 'Scientific Discovery and the Rate of Invention' in *The Rate and Direction of Inventive Activity*, op. cit.
48. W. Mueller 'Patents Research and Development and the Measurement of Inventive Activity' *Journal of Industrial Economics* (1966)

2 The Process of Invention

In this chapter one of the fundamental ingredients of technological change, invention, is examined in more detail. The first part of the chapter considers the nature of the invention process itself. The relationship between invention and economic stimuli is discussed in the second section. The third section then examines the patent system and its economic basis. The chapter is not intended to provide an exhaustive treatment of inventive activity; many other aspects of such activity are examined in other parts of the book where they are dealt with more appropriately.

2.1 The nature of invention

Invention must by definition be reflected in some novel feature in either a product or process. But novelty is clearly a term that embraces very different types of change. In some cases the novelty involved may appear small, for example a minor adjustment to an existing process at shop-floor level. Such an adjustment may be made by the production foreman as a normal part of his everyday duties. In other cases the novelty may be more fundamental in nature (even though its economic repercussions may still be small) and involve something more than just the exercise of acquired skills.

Some writers have attempted to make a clear distinction between these types of novelty. Usher for example distinguishes between novelty arising from 'acts of skill' and that arising from 'acts of insight'.[1] The former is 'a normal and continuous consequence of the skilled activities of engineers and technicians', undertaken within the environment of established processes and methods.[2] To Usher, the result of these acts does not constitute invention. Schmookler calls such changes 'sub invention'.[3] Where however a much higher level (in qualitative rather than quantitative terms) of novelty is introduced, which would not be the normal outcome of the application of skills previously acquired, an act of insight occurs. Such acts may arise out of the exercise of acquired skills with which they are often closely interlinked, but they involve much more. They are made possible 'by some special constellation of circumstances that invites or requires a special response'.[2]

Acts of insight however do not involve only major changes that are introduced very rarely by a few geniuses. To Usher they are characterised as 'numerous, pervasive and of small magnitudes'. Such acts of insight are the fundamental ingredient of invention. Of course, acts of skill and of insight are closely tied up with each other in the productive process, and at the margin their separation will depend on purely subjective judgements. Furthermore, acts of skill may require a high level of performance and may be just as important in economic terms as acts of insight. To some extent the distinction between these two kinds of acts finds its legal embodiment in patent laws in the provision found in many countries' legislation, that 'obviousness' is a ground for exclusion from patentability. Acts of skill, unlike those of insight, will normally be 'obvious' in the sense that they will usually be apparent to those who are 'skilled in the art'.

Many acts of insight occur while a firm is preparing for the commercial introduction of a new product or process. Teething troubles may require acts of both insight and skill for their solution. And, as shown in the previous chapter, major inventions may be usable only if further acts of insight occur. Because these activities are so much interwoven in this way, it is often very difficult to maintain Schumpeter's clear conceptual distinction between invention and innovation. The very process of commercialisation takes time and it is consequently rarely possible to pinpoint precisely when innovation in Schumpeter's analysis occurs (p. 262). Invention is often intextricably linked up with the process of innovating. Thus Ruttan suggests that invention is merely a special subset of technical innovation on which patents can be obtained, innovation being defined as any 'new thing'.[4] However there is often a fairly clear distinction between a 'primary' invention, which involves a radical departure from existing practice, and the inventions that are associated with its commercial introduction. It is the former type of invention that will usually be referred to in this book whenever invention and innovation are treated as separate activities. Such 'primary' inventions may not however occur as frequently as is often thought. For example, in his work on the development of the ship, which extended back to the use of the log as a means of water transport, Gilfillan could identify only a dozen 'distinctly original, abrupt and important' marine inventions. It is also worthy of note that not one of these 'ever revolutionised the ship, nor was itself entirely without precedent or starting points'.[5] And, as is shown clearly in part 3, it is a futile exercise to apportion 'credit' between those responsible for primary inventions and those responsible for secondary inventions.

Inventions do of course vary enormously in the extent of their departure from existing practice or from what was previously known. In the development of the hovercraft in this country (as shown in part 3) the basic patents taken out by Hovercraft Development Ltd on behalf of Cockerell reflected a far-reaching potential change in transportation techniques, although they did have some antecedents. Others have built on these patents, extending the air cushion principle to other fields, such as the movement of heavy loads, and improving the chances of commercial exploitation, eg by the addition and later the improvement of flexible skirts. It is interesting to note that the form of air cushion outlined in Cockerell's first patent in this field has been rendered out of date by further developments (see p. 253).

It is of course the primary inventions that receive widest acclaim in the literature, although the commercial application of these almost invariably involves further inventions. It is probably true to say, for example, that the hovercraft would not have become as much a commercial proposition as it is today without the introduction of flexible skirts. Such 'secondary' inventions usually involve much closer contact with the industrial environment. The common under-assessment of the frequent inventive acts of insight which are essential if a primary invention is to have practical use often results from an attempt to single out *one* hero or genius responsible for inventive activity in a particular field. While this may have the advantage of simplicity, it may provide a very distorted picture of reality, even though one might not go as far as Gilfillan's view that 'the popular notions of great inventors are essentially mythology'.[6]

32   SOME ECONOMIC ISSUES

Usher divides the inventive act (of whatever magnitude) into four distinct stages :[7]

1) the perception of a problem : typically an unfulfilled want;
2) the setting of the stage : the bringing together of the data and elements necessary for a satisfactory solution;
3) the primary act of insight : this stage, which provides the essential solution to the problem, is the most uncertain and its timing and nature may not therefore be predicted (as many case studies of invention have shown);
4) critical revision and development.

The first and last stages will frequently involve acts of skill, but all will incorporate acts of insight if important novel features are involved. For example the first stage may be impossible unless an invention has revealed the gap to be filled. The *economic* value of Usher's analysis lies in the fact that it identifies two stages (the second and the last) that can be consciously speeded up by the injection of funds. It is interesting to note here that most industrial R & D is aimed at *improving* products and processes, and at removing the 'bugs' in a development (see p. 167).

However, as indicated below, invention may not necessarily proceed along the paths laid down by Usher. For example his second stage may not be an essential prerequisite for his third step. Similarly, an invention may occur 'by accident' without the inventor actually perceiving a problem to be solved (stage one); indeed, an invention may create its own demand, rather than be a response to an 'unfulfilled want'.

2.2   Invention as an economic activity

None of the above discussion examines the question, 'What *causes* invention?'. Several schools of thought have emerged on this issue. One approach to the invention process is almost entirely negative : it holds that invention, by its very nature, cannot be explained. Invention is held to arise from 'the inspiration of the occasional genius who from time to time achieves a direct knowledge of essential truth through the exercise of intuition'.[8] Invention depends on, and is stimulated by, nothing. This approach has been rejected on historical grounds, and it is certainly not supported by Schmookler's evidence (see below).

Another school of thought, dominated by Ogburn and Gilfillan, has claimed that invention proceeds under the stress of necessity and that the individual inventor is therefore merely a tool of history :[9] Gilfillan for example has argued :

> There is no indication that any individual's genius has been necessary to any invention that has had any importance. To the historian and the social scientist the progress of invention appears impersonal.[10]

In this scheme of things invention is an inevitable responding to the needs of the time. Support for this thesis is sought through lists of 'duplicate' or equivalent inventions occurring at roughly the same time.[11] However, there are numerous examples of similar inventions occurring many years apart. Furthermore, as Schmookler has pointed out, many of these 'duplicate' inventions are distinct although they may be listed under one general heading.[12]

A third approach, closely linked with Usher's, is based on a hypothesis that sees major inventions resulting from a synthesis of previous minor inventions, each of which depends on an individual act of insight. Whereas the second approach outlined above concentrates on the demand side, this one gives rather greater emphasis to the supply side : all inventions depend first on perceiving a problem and second on the amassing of the necessary data.

Schmookler has attempted to test the importance of demand and supply forces in the invention process, and some of his evidence is discussed in a later section. Before looking at his work however, the measurement of inventive activity is examined. This issue is crucial to the quantitative testing of any hypothesis relating to inventive activity.

2.2.1 Measuring inventive activity

The only readily available measure of inventive activity is patent statistics. While these statistics have certain advantages – the inventions on which they are based will have been tested for a minimum level of technical soundness and novelty – there are a number of well-known limitations on their use in economic analysis.[13] The major drawbacks of such statistics for use in both cross-section and time-series analyses are briefly outlined below.

Firstly, these statistics give no measure of an invention's potential importance in either economic or technical terms. Even the minimum technical standards set by the patent authorities for an invention to qualify for a patent may change over time. Some inventions may open up a relatively new area, and have greater potential in terms of stimulating further inventions than others. Many inventions are never taken up commercially. Others, although taken up, do not become profitable. 'Taken up commercially' and 'profitable' are difficult terms to interpret, but there can be little doubt that inventions vary enormously in their economic effects. A weighting system would be necessary to overcome this defect, but to be fully effective it would require expert judgement on many difficult economic and technical matters. Which patents are important in a particular field is often hotly disputed by the companies and organisations involved (see p. 255). Sometimes the crucial issue of who owns the important patents can only be decided in the courts. The picture is further complicated for time-series studies by the fact that there may be a secular trend in the economic potential of inventions.[14] It should be remembered too that the 'economic importance' of an invention will depend also on the economic environment, particularly in relation to factor endowments in which it occurs, or is applied. For example, what constitutes an important invention in an advanced country may have little relevance in a developing economy.

One patent-weighting system which has been constructed by Bosworth is based on renewal data, the basic hypothesis being that the more commercially important a patent is considered to be, the more times it will be renewed during its life.[15] Thus, 'for any given year, the higher the average quality of inventive content, the greater the proportion of original patents that will be renewed during the subsequent sixteen years'. Bosworth has constructed an index of patent commercial importance based on this hypothesis. This index shows quite clearly that there has been a major increase in the commercial importance of patents since the interwar period. Although the use of renewal data in refining patent statistics has obvious value, its drawback is that it cannot distinguish between patents that are renewed for the same number of years.

Secondly, inventions may also differ between industries and over time in their technical complexity and in the resource inputs required to bring them to a patentable stage, irrespective of their technical or economic potential. Such variations may occur for instance because of different technical opportunities. Furthermore, much research work cannot in any case be patented, and inventions may differ in the amount of the non-patentable research required. Attempts to measure research productivity using patent data as an output measure must therefore be analysed with care. Inventions are also likely to vary in the amount of development work required to bring them to the commercial stage.

Thirdly, the propensity to patent inventions may vary between organisations and individuals, and over time; large corporations for a variety of reasons may be less inclined to patent than the small firm or individual. For the last-mentioned, the patent is often the only form of protection for his work (although it should be noted that if this protection is to be meaningful he must have the resources to maintain his rights in court). The large company on the other hand may be more ready to rely on the head start it obtains in launching a new product or process than on patent protection. Its existing production and marketing expertise, and its financial resources, may enable a quicker launch to be obtained. If this is so then it would be expected that as the structure of industry changes so patenting activity would also vary in a way unrelated to changes in inventive activity itself. Some industries, eg pharmaceuticals, are more likely to engage in patenting than other industries because of the ease with which products may be copied by existing or potential competitors.

These shortcomings in the use of patent data in the study of inventive activity are serious and have led some workers to abandon them. There are also a number of technical problems involved in the use of patents, which work against their use in economic analysis. For example, the patent classification system does not correspond very closely with the standard industrial classification. However, Schmookler has utilised patent data in his studies, arguing that 'we have a choice of using patent statistics cautiously and learning what we can from them, or not using them and learning nothing about what they alone can teach us'.[16]

2.2.2. *Supply and demand factors in inventive activity*

At the beginning of this section it was seen that both supply and demand pressures had been put forward as major determinants of inventive activity. On the one hand it is held that inventions will increase as the basic stock of knowledge increases. This knowledge may increase as a result of a scientific discovery, or of an invention that has combined existing knowledge in a new way, possibly with an additional input of new knowledge. On the other hand, invention may be viewed primarily as a response to a market opportunity or to a particular production problem. In popular jargon, necessity is seen as the mother of invention. These two approaches do of course overlap: for example, as already pointed out, a market opportunity or production problem may be apparent only after the acquisition of new knowledge. Nevertheless, it would be valuable to know the relative strengths of supply and demand forces.

Schmookler has made a useful contribution to this issue in his study of inventions in agriculture, petroleum refining, paper-making and railroads. He showed that, for the significant minority of cases in which the initiating stimulus for an important invention was identified, for almost all of these the stimulus was '*a technical problem or opportunity conceived by the inventor largely in economic terms . . . in no single instance, [was] a scientific discovery specified as the factor initiating an important invention in any of these four industries*' (Schmookler's italics).[17] While aware of the limitations of the methodolgy adopted to obtain the data, Schmookler was sufficiently confident to conclude that, while many important inventions depended on science, few if any were directly stimulated by specific scientific discoveries. In this way, society's stock of scientific knowledge is treated as a necessary condition for invention, but not as the prime stimulus. But as Kuznets has pointed out, 'many inventions involve, or quickly bring about, discovery of additional properties of the material universe'; ie invention may even *precede* scientific knowledge in a particular area.[18] (See also the footnote on p. 18).

Schmookler also took the view that, for some of the newer industries which he did not study, such as the electrical, electronics, pharmaceuticals and chemicals industries, his conclusions had to be only qualified rather than reversed. His reasons for this view were, firstly, that no relevant study reported antecedent scientific discoveries as an important determinant of corporate research in these industries; and, secondly, even if every radically new invention made in these fields was directly stimulated by some particular scientific discovery, 'it will still be true that the important inventions . . . which follow the basic one are likely to arise from the exposure of creative minds to technico-economic phenomena.'[19]

If Schookler's findings are accepted, what *does* determine the level of inventive activity? Again, Schmookler has suggested some tentative answers. He found that for the industries he studied (railroads, petroleum refining), capital goods inventions tended to vary directly with, and in response to, sales of capital goods in that field. This finding is supported by cross-section studies of other industries, which indicate that capital goods inventions tend to be distributed among industries in proportion to capital goods sales. These results seem to suggest that expected profitability and sales are important determinants of at least capital goods inventions. They also link in fairly neatly with other evidence which points to expected profitability as an important determinant of R & D expenditure (p. 164).

Schmookler's work however has been subjected to two main criticisms. Firstly, Jewkes *et al.* have argued that it over-emphasises the importance of commercial factors and does not pay enough attention to non-economic factors. They quote several examples where technology for technology's sake has been at work : for instance : 'Wankel became interested in the rotary engine because he considered "the shaking and pounding of the reciprocating piston engine unaesthetic as compared with the running of a turbine or an electric motor" '.[20] Such examples are useful to maintain a sense of balance, but it must be remembered that it is almost impossible to identify conclusively what the true motivating forces are, as opposed to what people may or may not say or genuinely think they are. (This difficulty also applies to Schmookler's analysis of the role of the scientific discovery in invention.) Furthermore, Schmookler's emphasis on demand factors is suggested by his statistical evidence, and is not dependent on an attempt to identify the motivating force directly. His work does not deny that, at the level of the individual inventor, important non-economic factors may be at work. However, taking inventive activity *as a whole,* it may nevertheless be possible to explain such activity largely in terms of economic stimuli.

The second criticism of Schmookler's work is more fundamental. In a recent article, Rosenberg argues that Schmookler's almost exclusive emphasis on demand factors as the determinant of inventive activity – which implies that the supply of inventions is perfectly elastic – ignores important variations in supply conditions between different fields.[21] The costs of and opportunities for inventing are very unlikely to be uniform across the whole of industry; indeed, in some areas (at any given point of time) invention simply may not be possible, despite clear evidence of demand pressure, because of the current state of the relevant science or technology. Rosenberg also points out that Schmookler's analysis has very limited explanatory value since his measures of demand are not defined *independently* of the evidence that the demand was met: 'In the absence of a reasonably clear, independent specification of the composition of demand, one can never demonstrate either that important components of demand have gone unsatisfied or that supply side factors played an important role in laying down the time pattern of inventive activity.'

2.2.3 *The exhaustion of inventive opportunities*

It may seem plausible to suggest that, as inventive activity in a given technological field increases, so the possibilities for further invention decrease as opportunities become exhausted. In other words, the greater the number of inventions produced, the nearer to perfection the available techniques become. Consequently the input per invention rises, while its value declines. Schmookler however argues that inventive activity declines not because of increasing costs of invention, but because of a fall in its value, owing to changed demand conditions. This is consistent with his analysis outlined above. In his interesting study of the horseshoe, for example, he shows that, despite the fact that the metal horseshoe was introduced in the second century B.C., the number of patents continued to rise in the United States until 1900, when the advent of other forms of motive power such as the internal combustion engine started to displace the horse. However, it must be asked *why* this displacement occurred. One very plausible answer would be that the opportunities for further development of 'horse technology' were *relatively* less abundant and attractive than those for improving other forms of motive power. This hypothesis would still be consistent with Schmookler's data on horseshoe patents.

2.3. The patent system

Most countries now have a patent system through which an inventor – an individual or an organisation – can obtain legally enforceable exclusive rights, subject to some qualifications (outlined later), to the use of his invention. In the UK these rights last for sixteen years from the date of the filing of a 'complete specification' and may under exceptional circumstances be extended to a maximum of twenty-six years. The patentee may license or withhold licences (again with certain limited qualifications) from any person he wishes.

A brief outline of the main characteristics of the patent system in the UK is given here. No attempt is made to criticise the patent-granting mechanism, or to become involved in detailed legal questions. A full appraisal is provided in the recent 'Bank's Report' on the patent system.[22]

When a patent application is made a specification must be filed with the Patent Office. This may be either a provisional or complete specification. A provisional application, which unlike its complete counterpart does not make claims for the invention but just describes it, must be followed by a complete specification within a maximum of fifteen months. The complete specification, besides making claims, must also fully describe the invention so that someone 'skilled in the art' is able to implement it. As Taylor and Silberston point out, however, many patents depend on other 'know-how' for their implementation, which for a variety of reasons is not included in the specification.[23] The advantage of a provisional specification is that it establishes a priority date for the invention if a patent is subsequently granted.

A patent will be granted in the UK if the invention, the claims for which are set out in the complete specification, has not been anticipated in an existing British patent, or in a prior publication, in the past fifty years. (In practice, the search is mainly restricted to patent specifications.) To this end the complete specification is examined and a search made of the relevant patents. No search of foreign patents is made, so that an applicant may be granted a patent on an invention that is already in use overseas. In the light of this examination the specification will usually be amended, so that only novel features are claimed; or in some cases, it will be abandoned. If the complete specification is accepted, it is then published and other parties have an opportunity to object, for example on the grounds that the invention is obvious. If no objections are made, or if they are overcome, then a patent is granted (this is usually about three years after the complete specification is filed). A patent may be revoked after granting on certain limited grounds (see below). Once a patent is granted fees must be paid each year after the first four years if the inventor's rights are to be maintained throughout the life of the patent.

An invention is defined in the 1949 Patent Act as 'any manner of new manufacture' or 'any new method or process of testing applicable to the improvement or control of manufacture'. Over 40,000 UK patents are granted annually, about seventy per cent of these coming from abroad. There are now over 200,000 patents in force in the UK.

The operation of the patent system outlined above must be seen in the context of certain important limitations and exceptions. Firstly, some products, processes or ideas are not patentable. For example, if the use of an invention would be contrary to law or morality it cannot be patented. Foodstuffs or medicines cannot be patented as such, although they may qualify as new chemical products. Theoretical ideas and scientific principles are also excluded.

Secondly, the Comptroller of Patents may grant a compulsory licence to a third party if he is satisfied that a patentee is abusing his monoply position. Section 37 of the 1949 Act sets out several grounds for granting such a licence, eg if demand for the patented article is not being met on reasonable terms, or if the invention is not being commercially worked, or worked to the 'fullest extent that is reasonably practicable' in the UK. In the special case of foods and medicines, the position is different and simpler: licences are granted on application unless it appears that 'there are good reasons for refusing the application'. Compulsory licences have rarely been applied for. This does not necessarily mean however, that this provision is unimportant. Companies may use the *threat* of an application for a compulsory licence to influence the policies of the patent holder.

Government departments and those authorised by them are in a special position. They may 'make, use and exercise', and where drugs and medicines are involved 'vend',* any patented invention for the services of the Crown (the inventor is however entitled to compensation).

On occasions, the National Health Service has used its powers to secure lower-priced drugs. For example, in recent years compulsory licences have been granted to two British companies so that they can produce the highly successful tranquillisers developed and produced by Roche under the brand names of Librium and Valium.[24]

It is instructive to note that these 'copying' firms have made very little headway in the tranquilliser market even though their products are a good deal cheaper. One reason for this lack of success may be that the brand names, which can only be used by Roche, are so firmly entrenched in the medical world that an outsider, especially a small one, faces very heavy distribution and marketing costs if he is to capture a substantial part of the market. Furthermore, Roche also provide the hospitals with tranquilliser supplies at greatly reduced prices; consequently when patients who have been given tranquillisers leave hospital and are taken under the care of their own general practitioners, there may be strong pressures on the latter to continue the treatment with the same branded products. In this case Roche has been able to build on its original patent position in such a way as to create a barrier to entry that cannot be surmounted solely by breaking through the patent barrier (see pp. 87–8).

Thirdly, the granting of a patent may be challenged by a third party within fifteen months of the publication of the complete specification. Hearings are by the Comptroller or by a senior patents examiner, with limited rights of appeal. Opposition to a patent at this stage may be based on criteria that cannot be used in the original examination of the specification by the Patent Office ('obviousness' is the main issue in most oppositions, yet this factor cannot be considered during the examination). The opposition may lead to revocation or amendment. Very few patents – probably less

* This last power is granted under Section 59, Health Services and Public Health Act, 1968.

than two per cent of accepted patent applications – are actually challenged;[25] again however the *threat* of such action still exists.

Finally, a patent dispute may be taken to the High Court on the initiative of the patent-holder where even wider tests of prior claim, obviousness, etc, may be applied, with considerable vigour. However the number of patent cases of this kind are very few. This may be partly due to industry's reluctance to use legal processes to settle disputes because of the costs involved, etc, but also because the strong patent which gives the patent-holder a clear stake in a particular territory will usually be enforceable without any recourse to legal processes. Weaker patents will be infringed anyway, and there will be a corresponding reluctance to attempt legal enforcement.

It is clear that the patent system is a complex mechanism. What are the *economic* arguments in its favour? Perhaps the most fundamental stems from the intrinsic nature of invention, which is the disclosure in some measure of some new process or product. If no patent system existed, so the argument goes, anybody would be free to copy, without charge, the ideas of a firm that has used its own resources to produce inventions. In welfare terms this would be a satisfactory outcome, since the information so produced is a public good : its stock does not decline as use increases. However, the incentive to undertake R & D might be less if firms carrying out such work know that the results obtained were available to all without charge, and that therefore there was no guarantee that they would obtain a sufficient return for the effort and risk-taking involved. Such a return may be available under the patent system one effect of which is to create property rights in knowledge. The return may come *either* from monopoly profits, when no other firms are licensed to produce the goods in question, *or* from licensing fees obtained from competitors who have access to the invention. A combination of both is often used with foreign firms obtaining licences to sell *outside* the home market for which few, if any, competing firms are granted licences.

The encouragement of inventive activity therefore is seen as a major argument for the patent system. However the objective of stimulating invention conflicts with the aim of stimulating maximum use of information *which has already been obtained* (see p. 125). This conflict has led some writers to propose a system in which the inventor does not have exclusive rights to his invention, but where the *state* provides him with a reward.[26] The reward given would however have to be far greater than a return for inventive activity alone, since the existing patent system may also act as an encouragement to new capital *investment* : if a firm knows it has protection for its product, it may be more willing to develop and lay down a production line for a new product, or to invest money in new machinery. (The patent system however cannot guarantee to provide *just* that level of profits necessary to induce a firm to undertake a particular activity or project; it may for example enable a firm to earn far higher profits than it would otherwise require on one project, but provide insufficient incentive in another). Another argument in favour of a patent system is that it encourages fuller disclosure at an early date, thereby reducing the likelihood of duplication of inventive activity.

How valid are these arguments? Firstly, it should be noted that even in the absence of a patent system companies may be able to rely on commercial secrecy to keep their technical lead. This lead may however be precarious where the mobility of scientists in an industry is high or where the invention is fully disclosed in the product (see chapter 5). Even in the latter case, however, duplication may be difficult if the imitating organisations do not have the requisite expertise.* Secondly, a firm may still be able to get a return for its work through 'know-how' agreements with other firms. These are usually closely linked with patent agreements, but in the absence of a patent system presumably could still be concluded. It should also be noted that in Britain at least firms have some limited common law rights in relation to the know-how they generate (see p. 104).

Thirdly, a firm producing an invention may rely on the 'lead time' it has over potential competitors in the market to recoup its outlays, fully aware that in the long run this lead will be eroded. However, this argument loses much of its force in the case of the

---

* In some cases, expertise may be required to assess whether or not the invention *is* fully disclosed in a product. One tragic example of where such expertise was required but was not recognised as being necessary is given in Nevil Shute's account of airship building after the First World War (*Slide Rule* London, Heinemann (1968)). Air Ministry officials designed and built the R38 airship simply by copying the size of the girders of the German Zeppelins. They made no attempt to calculate the aerodynamic forces acting on the ship. As a result of this empirical approach, the airship crashed with the loss of forty-four lives.

individual inventor who may not have the resources to set up in business. Patents may therefore be necessary for incentive purposes here. On the other hand, although *obtaining* a patent may be within reach of the small man, *enforcing* or *defending* patent rights may be beyond his resources.

Fourthly, the absence of patent protection is unlikely to prevent new inventions from occurring. Competitors are still likely to seek to improve their position by introducing new technical advances, although these advances may be of a different form. The curiosity-motivated private inventor is unlikely to curtail his activities as a result of the abandonment of a patent system.

Fifthly, it must be remembered that in the last analysis, the efficacy of a patent system depends on the ability of an inventor, individual or corporate, to *enforce* his (its) rights. Where costs of enforcement are high, the patent may lose some of its effectiveness and a firm may adopt alternative licensing strategies as a result. For example, in the US the Senate's Anti-trust Subcommittee's drug hearings heard the official of a medium-sized firm state that 'when they made discoveries . . . they no longer even bothered to try to market them as patented specialties, but rather just licensed them to all firms since the cost of bringing a patent infringement suit was simply prohibitive'.[27] If the existing patent system cannot be enforced anyway, then its demise would presumably not have far-reaching effects. However, as we have already pointed out, the *threat* of legal action may still be an effective method of enforcement.

Finally, a firm may rely on other non technological barriers, eg product differentiation, to maintain its competitive position (see section 4.3).

For all these reasons, it is clear that a good deal of R & D and inventive activity would still be carried out in the absence of a patent system, although, as Silberston has pointed out the direction of that research may change.[28] For example, there might be a tendency for the more risky projects to be abandoned. On the other hand there may be a *beneficial* shift of resources following the abandonment of a patent system, since the protection under such a system may lead to a greater concentration on patentable (as opposed to non-patentable) activities even though the former may not be more socially beneficial.

What the net effects of abandoning a patent system would be in practice cannot however be ascertained, since few 'before or after', 'with–without' comparisons can be made. At present the theoretical arguments for and against seem fairly evenly balanced. In these circumstances there would seem to be little justification for abandoning it altogether. Thus, as far as the system as a whole is concerned it seems that we must reluctantly accept Machlup's verdict :

> If one does not know whether a system 'as a whole' (in contrast to certain features in it) is good or bad, the safest 'policy conclusion' is to 'muddle through' either with it, if one has long lived with it, or without it, if one has lived without it. If we did not have a patent system, it would be irresponsible on the basis of our present knowledge of its economic consequences to recommend instituting one. But since we have had a patent system for a long time, it would be irresponsible on the basis of our present knowledge to recommend abolishing it.[29]

As Machlup implies however there may nevertheless be a case for piecemeal adjustments (see below).

The above discussion relates only to activities that may result in *patentable* products and processes. Chapter 1 described R & D activities that by their nature are unlikely to be patentable. It is very unlikely that the market will allocate an optimal amount of resources to these activities, even with a patent system. The role of government in this context is discussed in chapter 6.

2.3.1 Empirical studies of the patent system

Although the system has been the subject of considerable controversy among economists and has been the subject of an official committee of inquiry in the UK, very little empirical work has been done on its operation in advanced economies. Only one study of its economic effects has been done in the UK.[30] This study, undertaken by Taylor and Silberston, was aimed at studying the likely effects of introducing compulsory licensing. (Under the present system such licensing can be made outside food and medicines only if it can be shown that certain specific conditions are not being met.)

The findings of this study were based on an industrial inquiry which was carried out with the help of forty-four firms in a selection of research-based industries in manufacturing, which together account for most industrial patent activity in the UK. Firms were asked what difference it would have made to their operations if a system of 'worldwide thorough going compulsory licensing' had been in operation under which firms, although still able to patent, could not refuse applications for licences on commercially reasonable terms. This alternative to the present system was put forward on the grounds that it concentrated attention on the monopoly element in the system : 'the right of the patentee to refuse to permit others to use an invention that he wishes to work himself'.[31]

Taylor and Silberston's study is extremely comprehensive and covers a wide range of issues. Two of their findings are of especial interest to us here. Firstly, they found that only about eight per cent of the R & D carried out by the companies who provided them with usable returns was judged to be 'essentially dependent on patent protection'[32] (table 2.1). Most of this R & D was concentrated on pharmaceuticals and other finished and specialty chemicals. They also found, interestingly, that 'The small-scale efforts of the individual inventor and the very small firm do not seem to derive anything like as much stimulation from patent monopolies as has sometimes been thought'.[33] Secondly, they found

Table 2.1
Patent protection, R & D and production

Industry	R & D dependent on patent protection as percentage of respondents' R & D expenditure in the UK	Production dependent on patent protection as percentage of sales of UK production of respondents
Chemicals		
Pharmaceuticals	64	68
Other finished and specialty	25	5
Basic	5	negligible
Subtotal	17	10
Mechanical engineering		
Plant machinery and equipment	7	2
Components and materials	2	1
Subtotal	5	1½
Electrical engineering	negligible	1½
TOTAL	8	5

Note: The exact coverage of firms in the two columns is not the same. See the source text for further details.

Source: C. T. Taylor and Z. A. Silberston *The Economic Impact of the Patent System* Cambridge, Cambridge University Press (1973), tables 9.2 and 9.3

that roughly only five per cent of the sales of those companies providing answers would not have been produced if no effective patent protection had been available, ie if compulsory licences had been in operation. Again, 'patent-protected' sales are heavily concentrated in pharmaceutical and other finished and specialty chemicals.

The pharmaceutical industry particularly seems to be a special case in which the effect of patents may be considerable, even though the government already has considerable powers to grant compulsory licences. This in turn may spring from two special features associated with the pharmaceutical industry. Firstly, it is usually possible to make a very strong drug patent. Secondly, it may be very easy for a new entrant to set up production because of low scale economies and ease of manufacture. However, as we have seen, marketing eeconomies may still constitute a considerable entry barrier, even with ethical drugs.

The findings of this study must be treated with very great care. They are based on businessmen's stated reactions to hypothetical situations. (The authors are themselves very careful to stress the findings' limitations.) The possibility of bias in answers cannot therefore be excluded. It is possible for example that the answers given in relation to the extent of patent-protected production may reflect differences in profits obtainable from patent protection; it may be easier to obtain high profits from pharmaceutical patents. This in turn may be reflected in the responses given by interviewees to questions about their view of the importance of patents to production activities.

It must be remembered too that the study investigated the likely effects of a *marginal* change in the system : it did not indicate what might happen if the system was abandoned altogether. The existence of patent protection provides a way in which a return on expenditures may be obtained either through monopoly profit *or* through licensing. The imposition of compulsory licensing would still provide some return, whereas the absence of a system altogether would mean that a return was obtainable only from lead-times, industrial secrecy, non technological barriers to entry etc. It appears however that few firms in the UK at present attempt to obtain returns on their inventions through licensing other competitors in the domestic market (see p. 173).

One study in the United States, limited this time to the drug

48 SOME ECONOMIC ISSUES

industry, has compared the prices of certain drugs in six countries that allowed patent protection on pharmaceuticals with those in eleven countries where such protection is not granted.[34] Prices in countries operating a patent system were nearly forty per cent higher on average than in non-patent countries. If only developed countries are considered, the difference was sixty per cent. Blair comments on this study: 'There do not appear to be any factors other than the difference in patent policy that could explain price differences of this magnitude.'[35] The study also found that new discoveries were not dependent on the granting of a patent monopoly over a product. This conclusion is of particular interest in view of the findings of the UK study on the importance of patents in pharmaceuticals. The US study is of course limited to a sample of products within one industry. Nevertheless, as a substantial proportion of all UK patents are taken out in the pharmaceutical field, the study must carry some weight.

2.4 Conclusions

Inventive activity is clearly a complex activity for which there is no simple explanation. Inventions vary considerably in their technical complexity and potential. Measuring inventive activity creates considerable problems for the investigator: patent data, while very useful material, must be used with great caution. The existing studies on inventive activity using these data suggest that demand pressures are likely to play an important part in determining the pace in invention in different fields. Costs of inventing are likely to vary across industries however, because of differences in the current state of particular sciences and technologies. This in turn will also affect the rate of invention.

There are economic arguments both for and against the patent system. It is impossible however to come to a firm conclusion either way, at least within the limits of present knowledge. The one study carried out in the UK suggests that it may perform a useful role in stimulating inventive activity in certain sectors of industry, although this effect does not appear to be widespread. Another study concerned with the international comparison of the prices of drugs, where 'patent protected production' in the UK was found to be particularly important, suggested that in those countries where patents can be obtained on these products prices were much higher. It is unlikely however that this will apply across the whole of industry.

In the next two chapters we shall be concerned with the relationship between inventive and innovative activity and size of firm, taking the latter firstly in an absolute sense, and secondly in market terms.

NOTES on Chapter 2

1. A. P. Usher *A History of Mechanical Inventions* Cambridge, Mass., Harvard UP (1962) chapter IV
2. A. P. Usher 'Technical Change and Capital Formation' in National Bureau of Economic Research *Capital Foundation and Economic Growth* (1955) pp. 523–50, reprinted in N. Rosenberg (ed.) *The Economics of Technological Change* Harmondsworth, Penguin (1971) p. 47
3. J. Schmookler *Invention and Economic Growth* Cambridge, Mass., Harvard UP (1966) p. 6
4. V. Ruttan 'Usher and Schumpeter on Invention, Innovation and Change' *Quarterly Journal of Economics* (1959) reprinted in Rosenberg (ed.) op. cit. p. 83
5. S. C. Gilfillan *The Sociology of Invention* Follet (1935) pp. 64–5
6. ibid. p. 30
7. A. P. Usher (1962) op. cit. p. 65; also V. Ruttan op. cit. p. 79
8. A. P. Usher (1955) op. cit. p. 60
9. W. F. Ogburn *Social Change* Viking (1922) and S. C. Gilfillan (1935) op. cit.
10. S. C. Gilfillan (1935) op. cit. p. 10
11. W. F. Ogburn and D. S. Thomas 'Are Inventions Inevitable?' *Political Science Quarterly* (1922)
12. J. Schmookler op. cit. p. 191
13. For a more detailed discussion, see B. S. Sanders 'Difficulty in Measuring Inventive Activity' in National Bureau of Economic Research *The Rate and Direction of Inventive Activity* Princeton, NJ, Princeton UP (1962) and J. Schmookler op. cit. chap. II
14. S. Kuznets 'Inventive Activity: Problems of Definition and Measurement' in National Bureau of Economic Research (1962) op. cit. p. 38
15. D. L. Bosworth 'Changes in the Quality of Inventive Output and Patent Based Indices of Technological Change' *Bulletin of Economic Research* (1973)
16. J. Schmookler op. cit. p. 56
17. ibid. p. 67
18. S. Kuznets in National Bureau of Economic Research (1962) op. cit. p. 20
19. J. Schmookler op. cit. p. 69
20. J. Jewkes et al. *The Sources of Invention* 2nd edn London, Macmillan

(1969) p. 210
21. N. Rosenberg 'Science, Invention and Economic Growth' *Economic Journal* (1973)
22. *The British Patent System. Report of the Committee to Examine the Patent System and Patent Law* (Cmnd 4407) London, HMSO (1970). See also K. Boehm *The British Patent System:1. Administration* Cambridge, Cambridge UP (1967)
23. C. T. Taylor and Z. A. Silberston *The Economic Impact of the Patent System* (University of Cambridge Department of Applied Economics, Monograph 23) Cambridge, Cambridge UP (1973) p. 8
24. *A Report on the Supply of Chlordiazepoxide and Diazepam,* (Monopolies Commission, HC 197) London, HMSO (1973) chapter 3
25. C. T. Taylor and Z. A. Silberston op. cit. p. 13
26. For a discussion of the issues involved see K. Arrow 'Economic Welfare and the Allocation of Resources for Invention' in National Bureau of Economic Research (1962) op. cit.
27. Quoted in J. M. Blair *Economic Concentration* New York, Harcourt, Brace (1972) p. 207
28. A. Silberston 'The Patent System' *Lloyds Bank Review* (1967)
29. F. Machlup *An Economic Review of the Patent System* (85th Congress, 2nd Session, Senate Subcommittee on Patents Tradesmarks and Copyrights 1958 pp. 79–80) quoted in J. M. Blair op. cit.
30. C. T. Taylor and Z. A. Silberston op. cit. p. 388
31. ibid. p. 86
32. ibid. p. 198
33. ibid. p. 346
34. Quoted in J. M. Blair op. cit. pp. 388–91
35. ibid. p. 390

3 Size of Firm, Invention and Innovation

This chapter examines an issue that is important in terms of industrial policy: the relationship between firm size and technological change. In the first section the concentration of R & D spending in large companies is examined, together with a brief analysis of R & D productivity in firms of different sizes. Research intensities are also discussed. The second part of the chapter deals with the size of firm and invention and innovation.

3.1 Industrial R & D expenditure

3.1.1 *The concentration of R & D spending*

As with patent statistics, data on R & D spending contain pitfalls for the unwary. Firstly, they usually refer to formally organised R & D; consequently the firm undertaking 'informal' technical activities, eg on the shop floor, is not for statistical purposes pursuing R & D.[1] This probably leads to a bias against small firms in the data, since it is in these firms that relatively more informal R & D is carried out. Secondly, self-financed R & D by the individual inventor and the very small firm is excluded from the data. One estimate of the size of this omission puts it at about £75m. in the UK in 1967–68.[2] However, these shortcomings in the data are unlikely to affect the clear picture that emerges of the concentration of industrial R & D spending in a few large programmes. This is shown for five countries in table 3.1, which is taken from a study by the OECD. An earlier investigation (also by the OECD) in 1960 showed that 93 per cent of all industrial R & D in the UK was

Table 3.1
Percentage of total industrial R & D performed in firms ranked by size of R & D programme

Country	Year	4 largest firms	8 largest firms	20 largest firms	40 largest firms
USA	1963–64	22.0	35.0	57.0	70.0
UK	1964–65	25.6	34.0	47.2	57.9
France	1963	20.9	30.5	47.7	63.4
Japan	1963	—	—	—	47.7*
Italy	1963	46.4	56.3	70.4	81.6

* The first 54 firms

Source: *The Overall Level and Structure of R & D Effort in OECD Member Countries* Paris, OECD (1967) table 4

spent in firms of over 2000 employees (the figure for the US in 1958 was 90 per cent).[3] It is clear then that the majority of industrial R & D is carried out in large companies. Assuming that the statistics provide a reasonable picture of reality, is the evidence sufficient to support a policy aimed at the concentration of industry on the grounds that, if more large firms exist, R & D spending will rise?

3.1.2 *The threshold level*

Relevant to this question is the concept of a minimum *threshold* level of R & D activity, dictated by the technology of the industry, below which a company cannot carry out a viable R & D programme. B. R. Williams for instance suggested in 1962 that in some branches of chemistry a research team of less than twenty qualified research workers would not be worthwhile.[4] In 1965 C. Freeman gave figures of between £40,000 and £75,000 and between £2m. and £8m. for the threshold levels of annual R & D expenditure in radio communications receivers and communication satellites respectively.[5] There might therefore be a case for greater industrial concentration if it enabled more firms to reach the threshold level. This would then increase R & D spending overall.

A threshold may arise for two reasons. Firstly, the firm may be unable to remain in competition with existing firms unless it spends a given minimum on R & D. This implies that it will have to have a minimum level of sales so that its *unit* R & D costs do not put it in an uncompetitive position. Secondly (and linked with the first point), a threshold level may be imposed because of the intrinsic costs of conducting a particular research programme, eg the costs of indivisible equipment. Three qualifications must be made here however. Firstly, generalisations must be avoided; by the nature of the processes involved, and equipment required, high threshold levels may exist in certain industries, eg computers, but by no means in all. Secondly, there may be ways of mitigating the effects of threshold levels, for example by firms collaborating together and sharing equipment and thereby reducing costs (p. 198). Thirdly, there is little evidence that *beyond* the threshold there are continuing advantages of size; once a firm has reached the threshold level there may be no case for increasing industrial concentration still further on the grounds of R & D alone. There may however be other factors not directly related to R & D which favour a size larger than that necessary to support the threshold level of expenditure; for example, the bigger firm may be better placed to finance its capital programmes or conduct a marketing campaign. Conversely, the R & D threshold level may be so high that it creates *dis*economies in other areas of the firm's activities.

3.1.3 *The productivity of R & D and firm size*

The figures in table 3.1 refer only to *inputs* into the R & D process: what about the *output* per unit of R & D expenditure (ie R & D productivity) and firm size? The basic problem of measuring R & D productivity is of course one of providing a suitable output measure. Schmookler has used data on patents and has shown that in the six US industries for which data were available, R & D expenditure per patent pending in 1953 was much higher in the corporations with over 5000 employees than in those that employed fewer personnel.[6] This seems at first sight to suggest that the larger firm is less efficient in conducting its R & D. At the individual industry level Reekie has looked at the relative efficiency of R & D in different-sized firms in the pharmaceutical industry in Britain, measuring output of R & D by patents, over the past few years, and has concluded that: 'it would appear unwise to recommend the continued growth of the corporate giants' and that '. . . investors would be best recommended to invest in "small" companies, since in the immediate past they have had the best record of the technical progress on which commercial success in pharmaceuticals depends'.[7]

Reekie's conclusions *may* be right: the large firm may not provide as stimulating an environment for inventive work as the somewhat smaller firm. One reason for this difference may be that the smaller firm is more cost-conscious; tight budgets may in fact be conducive to creativity. For example, Jewkes *et al.* have commented: 'The fundamental discoveries that led to atomic energy were made by academic scientists with fairly simple tools. Perhaps the greatest of these scientists is reported as saying that "we could not afford elaborate equipment so we had to think" '.[8] The smaller firm which has a reputation for being innovative and adventurous may attract highly creative staff. The structure of its R & D department may also be less formal and more responsive to new, unorthodox ideas. However, as Schmookler points out, we cannot jump to such conclusions so rapidly – at least on the data that we have. Smaller firms' R & D expenditure may be under-represented in the data; the *nature* of the R & D undertaken by large firms may mean that it is less amenable to patenting (there is some evidence for this);[9] large firms may have a higher average quality of patent; and they may have a lower prospensity to patent.

Another possible measure of R & D output is innovation, although R & D is only one input into this process. Again, the evidence on the relationship between R & D productivity, measuring output by innovation, and firm size is very patchy. A recent British study covering over fifty per cent of manufacturing and construction net output suggests that the small firm of under 200 employees is more productive per unit of R & D expenditure than firms of over 200 employees.[10] An experimental weighting of the innovations did not substantially alter the picture. This may be partly explained by the fact that innovations often come into production via new firms which are necessarily small (see below). It may also result from a greater flexibility in small firms, and an often bigger dependence on innovation for commercial survival; this may give greater impetus to the new ideas being presented by scientists and engineers in smaller firms.

3.1.4 *Research intensities*

So far we have been considering R & D expenditures mainly in absolute terms. What is the evidence on the *relative* importance of R & D in firms of different sizes? To examine this issue we need some measure of research 'intensity'. Two measures have been used in the literature: firstly, *employment* in R & D as a percentage of total employment; and secondly, R & D *expenditure* as a percentage of sales or net output. Clearly, the indicator used will depend on the available data, and on the purpose to which the figures are to be put. The different measures may rank industries differently. For example, an employment measure of research intensity would give a higher figure in an industry whose productive capacity is very capital-intensive, than in an industry that had the same sized research establishment and the same output, but whose production activities were highly labour-intensive. But if expenditure data were used the industries might not differ so much. The use of sales figures (for which there is often no available alternative) has the disadvantages that the resulting ratio is heavily dependent on the degree of vertical integration of the industry or firm considered. The evidence on research intensities suggests that, whatever measures are used, there is a general, though by no means invariable, tendency for intensities to increase with firm size. This is shown in table 3.2, which gives data on total and company-financed R & D as a percentage of sales for three size classes in US manufacturing industry in 1965 (US figures are by far the most comprehensive).

There are however exceptions to the general rule of intensities increasing with firm size. For example, if we take company-financed R & D only, the medium-sized (1000–4999 employees) firms spend relatively more than the largest firms (more than 5000 employees) in drugs and medicines, other chemicals, primary and non-ferrous metals and scientific instruments.

Table 3.2
R & D intensities and firm size, US manufacturing industry, 1965

	Companies with total employment of:		
	less than 1000	1000–4999	5000 or more
Total R & D expenditure as percentage of net sales	1.8	2.3	5.0
Company-financed R & D expenditure as a percentage of net sales	1.4	1.5	2.1

Note: Later NSF surveys do not provide figures for the first two size categories; hence 1965 data are used here

Source: *Basic Research, Applied Research and Development 1965* Washington, National Science Foundation (1967)

The underlying pattern in research intensities in US industry shown in table 3.2 also exists in French industry.[11] In the UK the available evidence indicates that research intensities (measured by the number of QSEs (qualified scientists and engineers) in R & D per 1000 employees) is far lower for small firms (arbitrarily defined as firms with less than 200 employees) than for larger firms. Table 3.3 is taken from a study by Cox which covered about one-third of manufacturing employment in 1968.[12] The survey excluded firms with less than 11 employees. It should be noted that only one size division – at 200 employees – was made in this study: it is thus less comprehensive than the other surveys.

Table 3.3
QSEs in small firms in R & D compared with all firms in survey

	Number per 1000 employees	
	Small firms	All survey firms
All manufacturing	1.4	5.0
Food, drink, tobacco	0.4	1.9
Chemical and allied industries	7.9	22.1
Metal manufacturing	0.4	3.0
Mechanical engineering	1.4	2.8
Electrical engineering	1.8	5.5
Electronics	14.0	21.6
Aircraft	13.4	18.2
Motor vehicles	2.4	2.0
Other vehicles	1.7	4.6
Textile, clothing, etc	0.2	1.4
Other manufacturing	0.6	2.0

Source: J. G. Cox *Scientific and Engineering Manpower and Research in Small Firms* (Committee of Inquiry on Small Firms, Research Report No. 2) London, HMSO (1971) table 7

It can be seen that the only sector in which small firms have higher research intensities than the industry as a whole is motor vehicles. However, from other data, it is know that small firms in this industry employ relatively fewer QSEs *outside* R & D, eg in production and marketing.[13] This suggests that they are likely to be specialist companies producing research-intensive products, probably motor components. Small firms in electronics are also worthy of note since, although they have lower research intensities than large firms, their employment of QSEs *outside* of R & D more than outweighs this: QSEs in *all* activities, expressed as a percentage of all employees, is higher in small firms (5.8 per cent) than for the industry as a whole (4.4 per cent).[14] Electronics is the only industry in which this occurs. Cox suggests that this may be so because 'the technology of the electronics industry provides an exceptionally good environment for the employment of graduates in small businesses. A progressive and advanced group of technologies – making an appeal to talented graduates – can nevertheless be carried on and marketed with comparatively little expenditure on "hardware" '.[15]

All the studies on research intensities mentioned above classify firms into only a very few size categories. It is possible that *within* each of these size bands the *overall* figure disguises a downward trend in research intensities as firm size increases. This seems to be true at least for the largest size category (ie above 5000 employees). For example, Mansfield has shown that, in at least some American industries, the R & D-to-sales ratio is lower in the largest firms than in the somewhat smaller firms.[16] These findings are supported by Worley's more wide-ranging study of 198 very large firms in eight major US industries in which the relationship between the proportion of total employees engaged on R & D work and the size of firms was considered.[17] He concluded that : '. . . there appears a tendency for firms near the middle of the distribution to hire relatively more R & D personnel than do firms at either end, and in four of the eight industries, the firm employing the relatively greatest number of R & D personnel is smaller in size than the firm employing relatively the smallest number'. A similar pattern has been found by other investigators.[18]

These studies on research intensities may indicate that up to a certain size there are certain advantages to be obtained from a relative expansion of the research team, but that beyond that size the gains from further growth become progressively smaller. It must also be remembered that the R & D necessary for a product or product group is usually an initial fixed cost which is independent of the sales eventually achieved. Thus, in so far as large firms have bigger sales of any one product or range of products than somewhat smaller firms, they will also tend to have lower R & D-to-sales ratios. But if larger sales stem solely from greater *diversification* where the research programmes for each product are separate from each other, this argument would not apply. There may however be another explanation for the decline in research intensities at the top of the size spectrum. The move from the large to the very large size category may also imply a move from an oligopolistic environment to a position approaching monopoly. The former may, with its greater uncertainty induce higher relative R & D expenditures.

3.2 Size and technological change

3.2.1 *The advantage of size*

So far we have looked at the relationship between firm size and R & D expenditures. We now turn our attention to a more broadly based discussion of firm size and invention and innovation. Several economists have argued the case for size in the field of invention and innovation : only (they say) by the employment of large numbers of highly qualified men in expensive R & D laboratories, which only the large firm can support, and only by massive capital investment can modern inventions and ideas be conceived, translated into production and launched on to the market. Furthermore, it is argued that the large diversified firm (size appears to be closely linked with diversification)[19] is more likely to be in a position to utilise the 'surprise' results of R & D because of its bigger 'net' in the form of its existing product range. The high risks associated with technological innovations, and therefore the need to spread these risks by undertaking several technological developments simultaneously, gives further support to the argument for larger size. Large firms may also be able to obtain capital for such developments more easily and on more advantageous terms than their smaller counterparts. They can employ people qualified in many different disciplines, thus allowing cross-fertilisation of ideas. The more expensive aids to R & D can be purchased. They may also have the marketing and promotional framework for enabling an innovation to make a bigger and quicker impact on the market. These arguments are commonly used in the support of proposed mergers or rationalisation schemes.[20]

Several economists also hold the view, which in many ways is part of the argument for larger size, that much of inventive activity is now of a routine nature (see p. 5). Schumpeter, for example, argued that : 'Technological progress is increasingly becoming the business of teams of trained specialists who turn out what is required and make it work in predictable ways. The romance of earlier commercial adventure is rapidly wearing away, because so many more things can be strictly calculated that had of old to be visualised in a flash of genius.'[21] This sentiment finds its echo in the well-known quotation of J. K. Galbraith:

> a benign Providence ... has made the modern industry of a few large firms an excellent instrument for inducing technical change ...
> There is no more pleasant fiction than that technical change is the product of the matchless ingenuity of the small man forced by competition to employ his wits to better his neighbour. Unhappily it is a fiction. Technical development has long since become the preserve of the scientist and the engineer. Most of the cheap and simple inventions have, to put it bluntly and unpersuasively, been made.[22]

Galbraith stipulates not only size in absolute terms as a requirement for technological change, but also that the number of firms should be few, ie that they should be powerful in the market. In both these propositions (the second is discussed in the next chapter) he follows in Schumpeter's footsteps, although the latter's case for size rests on somewhat different foundations. The rest of this chapter is devoted to an examination of the existing evidence on the relationship between firm size and invention and innovation. Other arguments for or against large size, eg the need to achieve economies of scale in production, are not examined. These considerations may be more important than the ones discussed : we limited ourselves in this chapter however to the process of technological change.

3.2.2 *Invention*

It was shown earlier that inventive activity is not synonymous with R & D, at least as the latter is usually measured; it is necessary therefore to probe more deeply and to examine what relationship exists, if any, between the size of firms and their output of inventions. Unfortunately the existing evidence is rather disjointed and patchy, but the questions it raises have important policy implications.

One of the first studies in this field was published in 1958 by J. Jewkes *et al.* on the sources of sixty-one recent and important inventions (40 per cent were made after 1940). They concluded :

> More than one-half of the cases can be ranked as individual invention in the sense that much of the pioneering work was

carried through by men who were working on their own behalf without the backing of research institutions and usually with limited resources and assistance, or, where the inventors were employed in institutions, these institutions were . . . of such a kind that the individuals were autonomous, free to follow their own ideas without hindrance.[23]

Included in this category were automatic transmission, Bakelite, catalytic cracking of petroleum, streptomycin and Xerography. Among the remaining inventions are a number such as DDT, Cellophane and Terylene, which have arisen in relatively small firms. Thus, although several major inventions eg Nylon, the transistor and tetraethyl lead, have come from the research organisations of the large firms (about 20 per cent of the inventions studied), Jewkes's study has at least shown the continuing role of the individual and small firm in the invention process. Cheap and important inventions are still being made in considerable quantities. The second edition of the book, published in 1969, includes examples of more recent inventions supporting the original arguments. Many of these investigations have a tinge of the eccentric about them. For example, the dial telephone was invented by an undertaker and the hermetically sealed refrigerator by a French monk.[24]

Hamberg has supplemented Jewkes's results in his study of major inventions made during the decade 1946–55. Of twenty-seven inventions examined, only seven were the products of large industrial laboratories; the rest, including the oxygen converter, stereophonic sound and neomycin, came from independent inventors, small firms, universities and an agronomist more or less working for himself in an agricultural experimental station.[25]

It is of course difficult to know how representative the inventions chosen by Jewkes and Hamberg are of important modern inventions, particularly as the common *perception* of the degree of importance of an invention may not reflect its *actual* importance, as measured in some objective manner. For example, although the hovercraft is often acclaimed as an example of Britain's inventive genius, it is by no means certain that it will become an economically *successful* development (see part 3). This problem must be borne in mind in any interpretation of these findings.

Evidence relating to the inventive activities of different-sized firms in the upper size ranges only was presented by F. M. Scherer in 1965.[26] He examined the relationship between firm size and inventive output as measured by patent statistics for a sample of 448 firms, drawn from *Fortune's* list of the largest 500 American companies in 1955. He found that, although inventive output increased with firm sales, it did so generally at a less than proportionate rate. He also found that inventive output (and R & D output) did not appear to be systematically related to variations in market power. His findings 'raise doubts whether the big monopolistic, conglomerate corporation is as efficient an engine of technological change as disciples of Schumpeter . . . have supposed it to be' and suggests 'that a heavy burden of proof must be sustained by firms emphasising R & D potential as a justification (ie in merger cases) for bigness'. We should note however that Schumpeter's case for size related to *innovative* rather than inventive activity.

3.2.3 Some case studies

Several individual firm or industry studies have also been made in this field, a few of which are outlined here. W. F. Mueller has looked at the origins of the twenty-five process and product innovations introduced by Du Pont, one of the largest chemical firms in the world, over the period 1920–1950.[27] Of eighteen new product innovations, Du Pont *originated* only five and shared in the discovery of one other. Of seven process improvements Du Pont was much better represented: five were the results of its own research. Mueller's study suggests that Du Pont has been more successful in improving on inventions of others, than it has in providing its own. It is interesting to note too that the basic work on two of the three commercially important product innovations for which Du Pont was itself responsible – Neoprene and Nylon – was conducted in the 1920s and 1930s when the company's R & D budget was only a tiny fraction – even in real terms – of what it is today.

A study of the postwar aluminium industry has shown that the major aluminium producers were responsible for only a small proportion of new joining, fabricating and finishing techniques for aluminium over the period 1946–1957 – 11 per cent – compared with nearly 50 per cent which came from the relatively small equipment manufacturers.[28] They were responsible for only one out of the seven major inventions in these categories. On the other hand, the primary producers were responsible for 75 per cent of inventions in aluminium alloys, where, although the commercial returns may have been direct and immediate, they were not large: 'New alloys account for only a small fraction of the total sales of aluminium ingot . . . these inventions have apparently played a minor role in the increase in aluminium demand'. In the US petroleum refining industry all the (seven) basic 'cracking' inventions between 1913 and 1958 were the product of the individual inventor. The big refining companies played a major part in improvements, but the basic processes arose from external sources.[29] Finally, Hamberg's investigation into the origins of thirteen major innovations in the American steel industry between 1940 and 1955 (ten after 1950) showed that seven were invented by independent inventors.[30] It is interesting to note also that the role of the independent inventor appears to have increased over the period in this industry.

How should the findings of these individual studies be interpreted? It may be argued that they are specific to American experience and have little application elsewhere; that they relate to a few industries only; and that they involve several inventions produced either in the interwar period or immediately after the war, whereas the invention process in the 1950s and 1960s may have altered rapidly. But even after meeting these objections, it can at least be said that empirical support remains for the claim that the individual or small firm is still an important source of invention, even though the former may be declining in relative importance (see below). Galbraith's assertion therefore cannot go unchallenged.

On the other hand the studies do not provide sufficient justification for levelling accusations of lethargy and unimaginativeness at the large corporation. They may be lethargic or unimaginative but the fact that, in some industries at least, they are not the most important source of major inventions may just be a reflection of an economic division of labour in the invention–innovation process. This view ties in with the evidence (p. 167) that most industrial research laboratories *aim* at relatively short-term improvements, rather than at major inventions. For example, although Du Pont did not itself provide the bulk of the major inventions between 1920 and 1950, it nevertheless brought them into commercial production, ie acted as an innovator. The ability to utilise such external sources of invention requires an outward-looking attitude by the company, as well as a willingness to back high-risk investment. These requirements are often ignored in the literature.

Few of the studies mentioned above give any indication of changes in the different sources of invention over time. The only available breakdown that might be of use here is between individual and corporate invention. In the United States the percentage of patents taken out by individuals is decreasing (about 80 per cent of all patents in 1900; about 40 per cent in 1957), with corporations showing a corresponding increase.[31] Jewkes et al. confirm that this trend is broadly reflected in the UK statistics.[32] Adjustment for the lower propensity of larger corporations to patent would further emphasise this trend. It is probably also true that patents by individuals are relatively less likely to be economically important. It may therefore be true to say that, although the individual is still very important as a source of invention, his overall importance is declining. (Interestingly, Freeman suggests that the sample of inventions used by Jewkes et al. in the first edition of their book itself shows an increase in the relative importance of the very big corporations when comparing the inventions produced in the second part of the period studied, ie post 1928, with those produced in the first.[33] Jewkes et al. have however rejected this argument on the grounds of more recent evidence produced in their second edition).

3.2.4. The inventor's background

One of Galbraith's assertions is that invention is now – in contrast with the nineteenth century – the preserve of the trained scientist or engineer, for whom only the large firm can afford to pay. It is interesting therefore to note that in the United States a study has shown that over half of the recent inventors included in the sample were not 'college trained'.[34] The evidence is extremely limited but it does suggest a continuing role as an inventor for the person who is not a 'trained' scientist or engineer. It is also worth pointing out that an intimate acquaintance of the preceding technical and scientific literature may not be an essential prerequisite qualification for the inventor or scientist. Banting, the discoverer of insulin, was not acquainted with all the literature that preceded his work; indeed, if he had been he might have been put off from working in the field.[35] This of course is only one example limited to scientific investigation and it is far from providing evidence for a general case. However it does serve as a warning against oversimplified generalisations.

3.2.5 *Innovation*

The arguments for larger size appear much more persuasive when the innovation stage – the bringing into commercial production of an invention or idea – is considered. It is usually here (rather than in inventive activity) that risk capital in large quantities and the backing of a production and marketing organisation is required. R & D costs may in fact represent only a small percentage (say ten per cent) of the total costs of establishing a new good or process in production.[36] For instance, although Terylene was invented in a laboratory that was running on a budget of £200,000 p.a. (in 1940), £4m. was spent on pilot plant development and £15m. on new plant for the first major commercial production[37] (see also part 3). It would be dangerous however to generalise across industries. A further advantage of size is that it will enable the heavy fixed costs of setting up production to be spread more thinly over the output. It has been estimated for example that, for an aircraft such as the BAC 111, total unit costs could fall by up to a third as production extends from 30 to 100 aircraft.[38]

Before looking at the scanty evidence that exists on the relationship between size and innovation, the channels through which an innovation may come are described.

3.2.6 *Channels for innovation*

The ways in which an innovation may be produced are many and varied. In some cases the inventor also becomes the innovator by forming a new company. This is a relatively rare occurrence, for the simple reason that the qualities required of an inventor – imaginativeness, a non-conformist approach, an inquiring mind – are rarely combined, in the same person, with the commercial acumen and judgement required by the successful innovator. Furthermore, inventors are usually so convinced of the potential of the products of their own hands that they are unable or reluctant to face up to market realities. This inevitably makes possible sources of finance rather hesitant to give inventors a free hand to develop their products. This hesitancy is illustrated by a statement made to the author by a National Research Development Corporation official, who said that the Corporation had a golden rule : 'Never let an inventor near a pile of money'. Nevertheless, new companies have acted as major channels for innovation. A study of the development of radio, for instance, has shown that 'new companies have been of critical importance'.[39] The author's general comments on the role of the new firm in innovation of scientific advances into new products and new industries are interesting : 'It will be unfortunate if the translation of scientific advances into new products and new industries is left entirely to the great corporations. Any large, well-established institution almost inevitably tends to become bureaucratic.'

In his study of electrical engineering during the last quarter of the previous century, Passer has commented : 'As a general rule, the radically new developments came not from old, established firms, but from new firms and independent engineer-entrepreneurs.'[40] G. Brown, in his localised research on new firms established in Connecticut after the Second World War, found several started by inventors whose former employers were not interested in financing their inventions[41] (see also p. 102). There are numerous recent examples of new firms starting up in Britain on the basis of an innovation; for example, the plug board sequence control for machine tools, the printed circuit board for the electronics industry and atmospheric press packing for raw wool fibres were all important innovations introduced by new firms.[42]

In other cases, an innovation may be introduced by a large firm taking over a small firm, which has a viable new product or process which it does not have the resources to develop. Industrial boundaries drawn by statisticians do not of course necessarily impose any barrier to innovation. Several studies have shown how innovations in a particular industry have in fact come from outside that industry. The basic innovations in textiles for instance have come from the chemical industry. An Arthur D. Little survey[43] of the origins of recent major innovations in three 'mature' industries in the US – textiles, machine tools, and construction – has shown that the most important source of innovation has been the flow of technology across traditional industrial boundaries. In some instances an industry has borrowed a technology from another; in other instances firms in other industries have entered the industry, producing components, etc, and in yet other cases outsiders have manufactured a new version of their product. One of the most interesting recent studies showing relative outsiders acting as innovators in an established industry (aircraft) has concluded that,

'The innovators . . . shared one common feature: they were all comparative outsiders anxious to break into the market'.[44] Unexpected results of R & D may also lead a firm to diversify outside the industries in which it is currently operating.

3.2.7 Innovation and size

It is likely however that the bulk of innovation will be undertaken by firms already existing in the industry, although these firms may obtain the necessary technical knowledge from outside sources. The question then arises as to what relationship, if any, there is between firm size and innovation.

A contribution to the needed research has recently been made by Freeman, who analysed the role of firms of different sizes in innovation in the postwar period in UK industries accounting together for nearly fifty per cent of industry's total net output.[45] With the aid of independent advisers, a list of innovations (numbering 1100 in total) was drawn up for each industry together with the names of the firms originally responsible for their introduction. These firms were then contacted to confirm that they were the innovators, and to establish their size by number of employees at the time of the introduction of the new product or process. The results are given in table 3.4 and they are of course subject to the criticism acknowledged by Freeman that the innovations are not weighted by their importance. It may be argued however that the number of

Table 3.4
Number and percentage share of innovations by size of firm

Years	Small firms (1–199 employees)		Medium firms (200–99 employees)		Large firms (1000+ employees)		All firms	
	No.	% total	No.	% total	No.	% total	No.	% total
1945–53	17	9	25	12	160	79	202	100
1954–61	38	10	43	11	313	80	394	100
1962–70	54	11	53	10	399	79	506	100
1945–70	109	10	121	11	872	79	1102	100
Share of net output in 1958		21		79				100
Employment in 1958		25		75				100

Source: C. Freeman *The Role of Small Firms in Innovation in the UK Since 1945* (Committee of Inquiry on Small Firms, Research Report No. 6) London, HMSO (1971) table 2

innovations is sufficiently large for variations in importance to cancel each other out. (An experimental weighting system did not alter the results significantly.) It can be seen from the table that small firms were far less important as innovators than they were in net output or employment terms.

As expected, innovations in industries that were relatively capital-intensive (eg aerospace, motor vehicles, steel) and were therefore likely to require heavy investment were almost entirely produced by large firms. On the other hand small firms were relatively more important in industries where development and investment costs are typically lower, and where entry barriers are also low (eg scientific instruments, leather and footwear, construction).

In the United States, Mansfield has carried out a study of innovations in three basic industries, iron and steel, petroleum refining and coal.[46] While the results are admittedly rough, they are nevertheless worth considering. Mansfield set himself the task of establishing how many of the innovations between 1938 and 1958 were introduced by the four largest firms in each industry, and whether this was larger than their market share. He found that the largest four coal and petroleum firms were responsible for a larger number of innovations (weighted by importance) than their market share, but that the largest four steel companies produced fewer. His evidence fits in with the propositions that such firms would account for a larger share of innovations where the innovation requires a large investment relative to the size of the innovation's potential users, where the minimum size of firm to which the innovation would apply is large relative to the average size of firm in the industry, or where the average size of the largest four firms in the industry is much greater than the average size of all firms that are potential users of the innovation. In other words, the most desirable industrial structure in terms of innovation will depend on the nature of the innovation. In a similar study of the American railroad industry it was found from a limited sample that the four largest firms seemed to do a proportionately larger share of the innovating.

Thus, although these studies may provide a good deal of support for Schumpeter's hypothesis, the applicability of the latter will depend on specific industrial circumstances. There may however be a trend towards larger firms as relatively more important innovators: Mansfield showed that there are some grounds for saying that the smallest steel, petroleum and coal firms have done relatively less innovating in recent times than before the Second World War. Mansfield's results, as they stand, say very little about the relationship if any between the degree of competition in an industry and innovation. The possibility of such a relationship is discussed in the next chapter.

3.2.8 Successful v. unsuccessful innovations

It is probably true to say that nearly all the innovations analysed in the studies above were successful in the sense that they were commercially profitable. However, 'success' is a term of degree – some innovations turn out to be highly profitable, others less so. At the other extreme, many firms fail – again to varying degrees – in their attempts at innovation, even though they may aim at broadly the same market as the more successful firm. Is firm size an important ingredient of success? The Science Policy Research Unit at Sussex University has attempted to provide some pointers to the answer to this question in its SAPPHO (Scientific Activity Predictor from Patterns with Heuristic Origins) project.[47] For seventeen innovations in chemicals and twelve in instruments, the investigators paired a successful attempt at the innovation with an unsuccessful one. They then compared a large number of characteristics of both situations in an attempt to identify which factors were important in explaining success. They found that hardly any of the measures of size that they used were able to distinguish between success and failure. (There were exceptions. The size of the project team at the peak of R & D effort was significant. In chemicals, the size of team at the beginning of the project was also important; and in instruments, larger firms tended to be more successful if the parent company's employment was excluded.) Instead they found that:

> ... five underlying factors make up the majority of the success/failure differences. They are marketing performance (reflected in measures of publicity, sales effort, and user education), efficiency of development (reflected in measures of technical performance in eliminating 'bugs'), 'strength' of management and characteristics of managers (experience, power, responsibility of business innovators), effectiveness of communication, and understanding of user needs.[48]

These findings are in broad agreement with those by Carter and Williams (p. 186) and Langrish *et al.*[49]

The study at Sussex serves as a useful reminder of the dangers of undue concentration on 'single factor' explanations. However, the evidence would still be consistent with a relatively greater concentration of successful innovations occurring in larger-sized firms, and with the existence of a threshold level of R & D. (Nearly all the paired firms in the chemicals industry in the SAPPHO project had over 1000 employees.)

3.3 Conclusions

Most industrial R & D is concentrated in large firms. Research intensities tend to increase with firm size, although there is some evidence to suggest that this relationship does not hold beyond a fairly large size. Most of the empirical work that has been done on invention and innovation indicates that there is an important but possibly declining role for the individual and small firm, and that a number of important innovations have been made through the formation of new firms. This should serve as a warning against policies that aim to increase the rate of technological change by means of greater industrial concentration. Such policies may be helpful where they enable firms to reach a threshold level of R & D spending. Beyond this level however there may be little to be gained from greater size, at least as far as R & D activity is concerned. As Scherer points out : 'No single firm size is uniquely conducive to technological progress. There is room for firms of all sizes. What we want, therefore, may be a diversity of sizes, each with its own special advantages and disadvantages.'[50] Scherer's view is clearly supported by the results of the Sussex study. Successful innovation depends on numerous factors, of which size of firm may be only one.

NOTES on Chapter 3

1. For a discussion of the limitation of official R & D statistics, see B. S. Saunders 'Some Difficulties in Measuring the Level of Inventive Activity' in National Bureau of Economic Research, *The Rate and Direction of Inventive Activity* Princeton, Princeton UP (1962)

2. C. T. Taylor and Z. A. Silberston *The Economic Impact of Patents* Cambridge, Cambridge UP (1973) pp. 57–8
3. *Science and Economic Growth and Government Policy* Paris, OECD (1963) p. 87
4. B. R. Williams *Technology, Investment and Growth* London, Chapman and Hall (1967) p. 85
5. C. Freeman 'Research and Development in Electronic Capital Goods' *National Institute Economic Review* (1965)
6. J. Schmookler *Invention and Economic Growth* Cambridge, Mass., Harvard UP (1966) p. 33
7. W. D. Reekie 'Location and Relative Efficiency of R & D in the Pharmaceutical Industry' *Business Ratios* (1969)
8. J. Jewkes et al. *The Sources of Invention* 2nd edn, London, Macmillan (1969) p. 68
9. C. T. Taylor and Z. A. Silberston op. cit. p. 150
10. C. Freeman *The Role of Small Firms in Innovation in the United Kingdom since 1945* (Committee of Inquiry on Small Firms, Research Report No. 6) London, HMSO (1971)
11. J. C. Morand 'La recherche et le developpement selon la Dimension des enterprises' *Le Progres scientifique* (1968)
12. J. G. Cox *Scientific and Engineering Manpower and Research in Small Firms* (Committee of Inquiry on Small Firms, Research Report No. 2) London, HMSO (1971)
13. ibid. table 2
14. ibid.
15. ibid. p. 5
16. E. Mansfield *The Economics of Technological Change* New York, Norton (1968) p. 94
17. J. S. Worley 'Industrial Research and the New Competition' *Journal of Political Economy* (1961)
18. Eg W. Comanor 'Market Structure, Product Differentiation and Industrial Research' *Quarterly Journal of Economics* (1967); D. Hamberg 'Size of Firm Oligopoly and Research: The Evidence' *Canadian Journal of Political Economy* (1964)
19. M. Gort *Diversification and Integration in American Industry* National Bureau of Economic Research, Princeton, Princeton UP (1962)
20. P. S. Johnson 'Firm Size and Technological Change' *Moorgate and Wall Street* (1970)
21. J. A. Schumpeter *Capitalism, Socialism and Democracy* 5th edn, London, Allen & Unwin (1952) p. 132
22. J. K. Galbraith *American Capitalism* Harmondsworth, Penguin (1956) p. 100
23. J. Jewkes et al. op. cit. p. 73
24. Quoted in G. Bannock *The Juggernauts* London, Weidenfeld and Nicolson (1971) p. 178
25. D. Hamberg 'Invention in the Industrial Research Laboratory' *Journal of Political Economy* (1963)
26. F. M. Scherer 'Firm Size, Market Structure and the Output of Patented Inventions' *American Economic Review* (1965)
27. W. F. Mueller 'The Origins of the Basic Inventions Underlying Du Pont's Major Product and Process Innovations, 1920–1950' in National Bureau of Economic Research op. cit.
28. M. J. Peck 'Inventions in the Post War Aluminium Industry' in National Bureau of Economic Research op. cit.

29. J. Enos 'Invention and Innovation in the Petroleum Refining Industry' in National Bureau of Economic Research op. cit.
30. D. Hamberg op. cit.
31. E. Mansfield op. cit. p. 91
32. J. Jewkes et al. op. cit. p. 89
33. Quoted in Jewkes et al. op. cit. p. 208
34. J. Schmookler 'Inventors Past and Present' *Review of Economics and Statistics* (1957)
35. J. Jewkes et al. op. cit. pp. 260–1
36. B. Lloyd 'Invention, Innovation and Size' *Moorgate and Wall Street* (1970)
37. J. Jewkes et al. op. cit. p. 312
38. S. G. Sturmey 'Cost Curves and Pricing in Aircraft Production' *Economic Journal* (1964)
39. W. R. Maclaurin *Invention and Innovation in the Radio Industry* New York, Macmillan (1949) pp. 243–4
40. H. C. Passer *The Electrical Manufacturers 1875–1900* Cambridge, Mass., Harvard UP (1953) p. 358
41. G. Brown 'Characteristics of New Enterprises' *New England Business Review* (1957)
42. C. Freeman (1971) op. cit.
43. Quoted in E. Mansfield *op. cit.* p. 111
44. R. Miller and D. Sawers *The Technical Development of Modern Aviation* London, Routledge & Kegan Paul (1968) p. 263
45. C. Freeman (1971) op. cit.
46. E. Mansfield op. cit. p. 108
47. Science Policy Research Unit *Success and Failure in Industrial Innovation* London, Centre for the Study of Industrial Innovation (1972)
48. ibid. pp. 16–17
49. J. Langrish et al. *Wealth from Knowledge* London, Macmillan (1972)
50. F. M. Scherer *Industrial Market Structure and Economic Performance* New York, Rand McNally (1971) p. 357

4 Competition in Innovation

As we have already noted (p. 15), a substantial proportion of micro theory in economics is built on static models which are constructed in the context of a *given technology*. Competition between firms is seen principally in terms of changes in prices and/or product differentiation activities. Schumpeter, however, took a much more fundamental view of competition. He argued that in the context of his capitalist engine of 'creative destruction', in which innovation is continually destroying the old and building up the new, the competition that counts is that from 'the new commodity, the new technology, the new source of supply, the new type of organisation'. This type of competition 'strikes not at the margins of the profits and the outputs of the existing firms but at their foundations and their very lives'.[1] No producer, not even the entrenched monopolist, can protect himself from this threat in the long run. To Schumpeter the existence of such competition, which is continually threatening established positions, emphasised the importance of looking at industrial operations not at a given point in time, but in the context of a much longer period.

As indicated in the previous chapter, innovations do not necessarily come from inside traditional industrial boundaries. For example, glass container manufacturers face competition from the plastic materials developed in the chemicals industry; and in textiles the advent of man-made fibres has accentuated the decline of manufacturers who base their operations on natural fibres. Challenge through innovation may also come from entirely new firms entering the industry or from established outsiders. The history of the radio and television industry provides some good examples of the former kind of challenge.[2] When point-to-point radio communication was first introduced between Britain and the US and other countries by Marconi, for example, the established cable companies that operated agreements on rates refused to take the threat seriously, and only after a lapse of several years did they attempt to improve their own methods of transmission to meet the new challenge. (Earlier, however, they had attempted to prevent Marconi's entry into long-distance communication.) Later, Marconi in turn had its strong monopoly of marine radio communications threatened by the entry of new rivals, all of whom based their attacks on technical advances. After the Second World War it was the turn of the existing radio industry to be threatened by another major innovation, television, which was introduced, by yet another newcomer, J. Baird.

While Schumpeter took the view that no monopoly could be maintained in the long run in a business environment that was dominated by innovation, he also argued that *existing* market power where firms acting singly or in concert could raise prices above, and restrict output below, the competitive level, was in fact *conducive* to technological change. The rest of this chapter will be devoted to a discussion of this and related hypotheses. The first section reviews the various theoretical arguments; the second section concentrates on the empirical evidence; and the final section deals with technological entry barriers.

4.1 Market structure and innovation: the theoretical arguments

The arguments for market power in relation to technological change take two main forms. Firstly, some economists, notably Schumpeter, have argued that such power may be a necessary *condition* for innovation. The causation is seen to run from market power to innovation, with the existence of an established position of strength favouring innovation. Secondly, the prospect of *future* market power obtained from an innovation that is protected by patents and industrial secrecy acts as an *incentive* to innovation. Against these two may be placed the argument that competition between firms creates *pressures* for innovation. These viewpoints are examined in turn below. All may be true to some extent. But if recommendations on industrial structure are to be made, it is necessary to find out *how* important each factor is, and whether a particular structure that maximises the rate of innovation in a particular industrial environment can be identified. (As indicated later however, even if such a structure could be identified, it does not necessarily follow that it is optimal *overall,* because other considerations, eg production economies, have to be taken into account. Furthermore, the structure of a particular industry will determine the extent of competitive duplication of R & D. Thus, even if it favours innovation, the costs of such a structure in terms of duplication may outweigh the benefits).

4.1.1 *Existing monopoly leads to innovation*

In some cases a concentrated market structure may be necessary to achieve the threshold level of R & D discussed in the previous chapter and economies of scale in other spheres, given the existing size of the market. This possibility is not discussed in detail here, however.

Schumpeter based his case for market power on two main grounds. His first argument rests on the fact that innovations necessarily involve change – often of a radical nature – in production techniques, organisation and marketing. In this context, he saw restrictive practices or the monopolistic position as providing temporary security and certainty, so that the firms involved could aim more surely at the 'indistinct and moving' target of innovation. One way in which this increased certainty may help, for instance, is in the planning of R & D projects which may take some time to reach fruition. (Schumpeter did however agree that *some* restrictive practices might not be beneficial in the long run, although he failed to specify which agreements might come within this category).[3] Schumpeter's second argument in favour of a strong market position in the innovative process was that the firm or firm's involved are likely to be more profitable than their competitive counterparts. Hence their access to funds, especially internal funds, for investment in development work and the setting up of productive capacity based on new technology will be greater; thus, a restrictive strategy 'might still prove to be the easiest and most effective way of collecting the means by which to finance additional investment'.[4]

The evidence on this second issue is rather thin and ambiguous. Scherer examined the relationship between company profits (as a percentage of sales) and liquid assets (as a percentage of total assets) – both at 1955 – and 1959 patents per billion dollars of 1955 sales for the largest US firms.[5] In both cases the correlations he found were not significantly different from zero. Minasian's study of US chemicals (p. 13) also support's Scherer's findings. Grabowski's investigation into three US industries (p. 164), on the other hand, shows that internally generated funds are an important determinant (along with others) of R & D expenditures. These studies use different data and they relate in any case only to R & D inputs and outputs with the latter measured by patents. They do not relate to physical investment, which is likely to play a very important part in innovative activity. It must be remembered too that a strong market position may not always show itself up in higher profits, yet it may still act as a good environment for innovation. This is because what would otherwise be shown up as higher profits may be used to pay higher salaries and to provide better working conditions (ie to increase costs), which in turn attract better R & D and other personnel.

The implication of Schumpeter's analysis is that there may be a conflict between a market structure that is statically efficient in the allocative sense and one that is dynamically efficient in terms of maximising the rate of innovation. In Schumpeter's own words, 'a system . . . that at *every* given point of time fully utilises its possibilities to the best advantage, may yet in the long run be inferior to a system that does so at *no* given point of time, because the latter's failure to do so may be a condition for the level or speed of long run performance'[6] (Schumpeter's italics). Thus, while perfect competition, for instance, may ensure allocative efficiency in a static Paretian sense it is, as Cyert and George point out, neither a necessary nor a sufficient condition for innovation.[7]

It should be noted however that Schumpeter's hypothesis as originally stated does not necessarily imply that innovative activity should be a continuous and increasing function of market power.[8] A *threshold* level of market power would be consistent with his arguments.

Another argument in favour of market strength is that the firm concerned may be able to capture a greater part of the total returns from an innovation that can be easily copied than could a firm in a highly competitive industry, which might find its innovation being taken up by all the other firms in the industry. On the other hand, if an innovation is *difficult* to imitate, the firm in the competitive industry has more to gain, in terms of market share, than has a monopolist or oligopolist.

76 SOME ECONOMIC ISSUES

In contrast to Schumpeter, Arrow has built up a theoretical case for arguing that the monopolist's incentive to invent* – as measured by how profitable an innovation is likely to be – is less than that for an inventor licensing his invention to a competitive industry.[9] Arrow defines a competitive situation as one in which 'the industry produces under competitive conditions while the inventor can set an arbitrary royalty for the use of his invention', whereas 'in the monopolistic situation, only the monopolist itself can invent'. In both situations, producers are assumed to be operating under the same constant costs. For the sake of exposition, Arrow also assumes a cost-reducing invention, although the case of a new product invention would not fundamentally alter the analysis.†

In figure 4.1, unit costs (which are assumed to be constant) before the invention are c and the marginal revenue and demand schedules MR and D respectively. This would lead to an equilibrium output of X_c under competition and X_m under monopoly. Price under competition would be c and under monopoly P_m. Now let us suppose that a cost-reducing invention occurs and costs fall to c'. In figure 4.1(a), the cost reduction is such that under the monopoly situation the new price (P'_m) would still be above the old competitive price (c). The new competitive price and output, with no royalties being charged, would be c' and X'_c respectively. In this competitive situation, however, the inventor could charge unit royalties of up to $c - c'$. (He cannot charge more as firms would still be able to utilise the old methods of production.) The maximum total royalties obtainable, ie the 'incentive to invent', would therefore be $X_c(c - c')$. Under monopoly, however, the incentive to invent is represented by the shaded area, ie the change in total revenue, as measured by the increase in the sum of the marginal revenues, less the change in total costs.‡ This will always be smaller than $X_c(c - c')$.

* This is the incentive to invent *given* a particular market structure. As such, Arrow's analysis counters Schumpeter's argument for monopoly as a condition for innovation, and is distinct from the 'future monopoly power is an incentive to innovate' argument.
† It seems as if Arrow is using the term 'invention' as synonymous with innovation in the Schumpeterian sense.
‡ In Arrow's terminology, the incentive to invent under monopoly is

$$P' - P = \int_{X_m}^{X'_m} R(x)dx - c'X'_m + cX_m$$

where P' is post- and P is pre-invention profits and $R(x)$ is marginal revenue.

COMPETITION IN INNOVATION 77

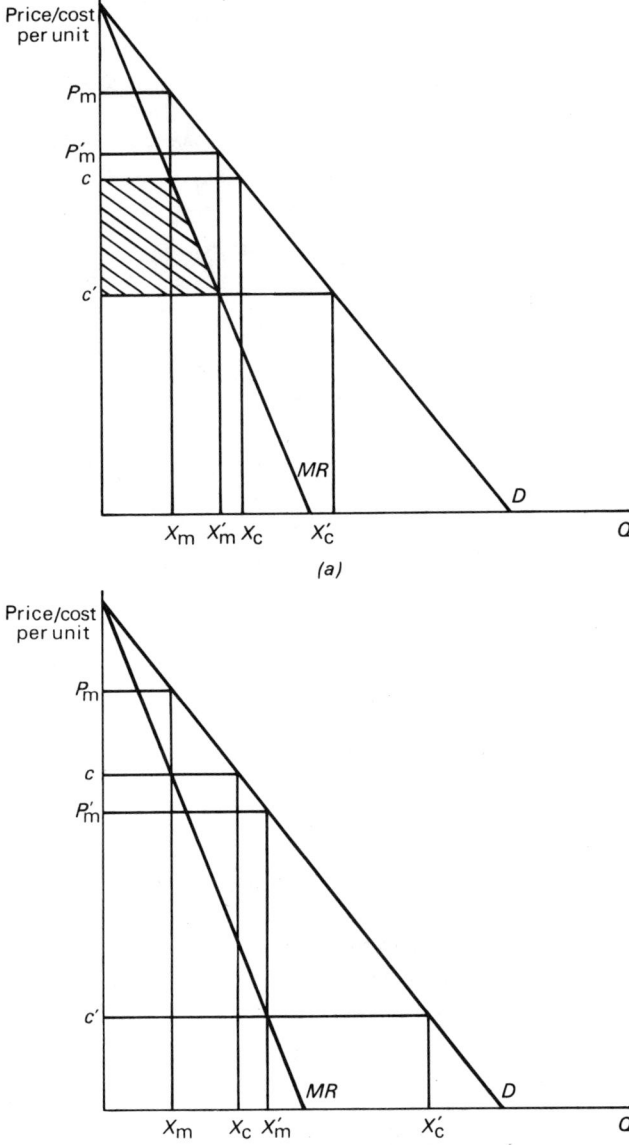

FIGURE 4.1(a) and 4.1(b)

In figure 4.1(b) different assumptions are made about the extent of the cost reduction made possible by the invention. Here it is sufficiently large to reduce the new monopoly price (P'_m) to *below* the old competitive price (c). The inventor in the competitive case will then find that his most profitable unit royalty is $P'_m - c'$, giving him a total profit of $X'_m(P'_m - c')$. This is the same level of profits that a monopolist would earn with the same cost-reducing invention. But because the monopolist is *already* earning higher profits than he would have done under competition prior to the invention, ie $X_m(P_m - c)$, his incentive to produce the new invention is inevitably less. It is in fact:

$$X'_m(P'_m - c') - X_m(P_m - c)$$

Demsetz has however challenged the conclusion reached in Arrow's analysis.[10] He compares the situation in which the inventor sells his invention to a competitive industry with that in which the invention is sold to a monopolist, in both cases receiving a royalty in return. No longer is the monopolist the inventor. He argues that, if the inventor is restricted (for instance by legal factors) to charging both industries the same royalty, and if allowance is made for the fact that the monopolist will *in any case* normally produce less output than a firm in a competitive industry, then Arrow's results do not hold.

Demsetz's argument can be illustrated through figure 4.1(b). Suppose that identical unit royalties $P'_m - c'$ are charged to both industries. The competitive industry will pay total royalties of $X'_m(P'_m - c')$ while the monopolist will pay only half this amount, because he restricts output to where MR intersects P'_m. If, however, the effects of different outputs in competitive and monopolistic industries are neutralised by defining MR in figure 4.1(b) as also being the demand curve of the competitive industry (so that, for any level of costs, the *same* level of output is produced), then it is clear that the inventor will receive the *same* royalty from both industries:

$$\tfrac{1}{2}\,[X'_m(P'_m - c')]$$

Demsetz's conclusion therefore is that, once these adjustments are made, 'there is no special adverse effect of monopoly on the incentive to invention'.

Furthermore, in the case in which royalties do not have to be charged at the same rate, Demsetz demonstrates that in equal-sized industries, the monopoly incentive is *greater* than the competitive counterpart. In figure 4.2 the competitive industry's demand curve (D_c) is again assumed to be the same as the marginal revenue curve of the monopolist (MR_m). MR_c is the marginal revenue curve based D_c. With pre-invention unit costs at c, competitive and monopoly outputs are the same: $X_c = X_m$ (at prices c and P_m respectively). The post invention cost level is c'. As in figure 4.1(b), the incentive for the monopolist's invention is:

$$X'_m(P'_m - c') - X_m(P_m - c).$$

In the competitive industry, the maximum per unit royalty the inventor can obtain is $P'_m - c'$. This will maximise the inventor's royalty income $X_i(P'_m - c')$. It is clear from visual inspection of the figure that this is smaller than the monopolist's incentive. Demsetz has proved algebraically that this is a general conclusion for the linear model. His conclusions are of considerable interest,

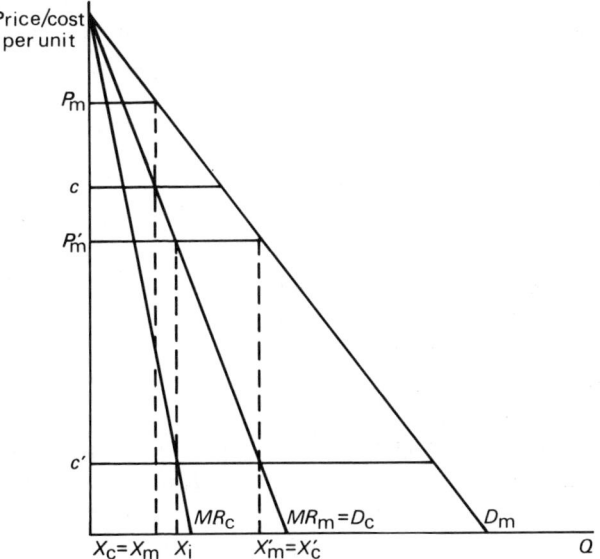

FIGURE 4.2

especially when examined in the context of Schumpeter's hypothesis outlined earlier, since the analysis suggests that 'at least in the linear model of two industries of equal output size, the more monopolistic will give the greatest encouragement to invention', even though on other efficiency grounds an anti-trust policy might be desirable.

Both Arrow and Demsetz however ignore important institutional factors. Yamey, for example, has pointed out that the situation in which the inventor sells to the monopolist is one of bilateral monopoly which 'introduces the element of indeterminacy and bargaining into the pricing process'.[11] This is likely to be particularly important where bargaining is conducted *after* the inventor has committed his resources, where he would therefore be in danger of exploitation. Even where a payment is negotiated before an invention is made, the actual amount will depend on the negotiating skill of the parties. This will be particularly true where the outcome of an invention, in terms of cost reduction, quality improvement, investment needed to introduce it commercially etc, is difficult to determine *ex ante*, as is usually the case. Furthermore, the inventor will be dependent on a sole customer for the market assessment of his invention. This further increases the risks faced by the inventor. Arrow did, of course, assume that only the monopolist invented. This overcomes the above problems to some extent, although sources of market information even for the monopolist inventor are still limited.

4.1.2 *Future monopoly position as an incentive to innovate*

Higher profits that come from establishing a monopoly position may provide a strong incentive to innovation. By successful innovation, the firm may be able to obtain a technological lead which can be protected – at least for a period – in various ways, eg by patents, industrial secrecy etc. The monopoly carrot for the inventor is of course the basic economic rationale behind the patent system. We shall see, too (p. 164), that expected profitability may be an important determinant of R & D budgets at the firm level. On the other hand other factors, eg the desire for the quiet life or for prestige, are also likely to play some part in influencing the pace of innovation.

4.1.3 *Pressure to innovate from competition*

The essence of this argument is that intensive price competition between existing manufacturers may increase the pressure on firms to innovate, whereas monopoly may lead to firms becoming lethargic in the innovation process. Cyert and March, for example, argue that competitive pressures may lead firms to start a 'search' process for lower costs.[12] This pressure will be particularly strong for the less efficient firm, which is likely to be swallowed up or put out of business by its more efficient brethren unless it can innovate. Downie saw the process of innovation by such firms as checking the 'transfer mechanism' by which efficient firms expand at the expense of the less efficient.[13]

Theoretical predictions about the relationship between market structure and innovation are clearly not unambiguous and will depend, as is so often the case, on the particular assumptions adopted. Industrial conditions, profit incentives and competitive pressures are all likely to play their part. In some cases, they may push (or pull) in opposite directions. For example, intense price competition may lead a firm to seek an improvement in its position via innovation, but the low profits that may exist in such a situation may deny the firm access to the necessary funds.

We now turn to the empirical evidence on the relationship between market structure and innovation.

4.2 Market structure and innovation: the empirical evidence

There is now a considerable body of empirical material which is of direct relevance to the discussion of the relationship between market structure and innovation. Much of it is of a case study nature. One of the earliest investigations of this kind was by Bright and Maclaurin on the introduction of the fluorescent lamp in the United States.[14] General Electric (GE) was the dominant producer of incandescent lamps, with nearly sixty per cent of the market for large, tungsten filament lamps (the most important market). Its R & D effort dominated the market: none of the smaller firms had the resources to provide a serious challenge by producing new or improved products and processes. All the other important firms were licensed from it and their sales were limited by quotas. GE also had agreements with foreign producers to restrict the latter's sales in US markets. As a consequence of GE's entrenched position, 'the development of radically new types of lighting devices tended to take pace slowly'.

The implication of this study – that market dominance impedes innovation – is supported by Miller and Sawers in their investigation into the technical development of aircraft. They concluded that 'the clearest lesson' one can draw from the history of the aircraft industry is that 'competition stimulates innovation'.[15] Both Douglas and Lockheed, each with a substantial share of the US commercial market, adopted very much a passive role to new innovations in the later 1940s and 1950s. However, when the Comet and later the 707 appeared – both aircraft produced by relative outsiders, with little stake in the existing commercial market – they were then forced to enter jet production themselves. The study by Peck of the US aluminium industry (p. 61) provides another example of an unconcentrated industry (equipment manufacturing) providing a stream of important innovations.

On the other hand, numerous examples exists of technological change occurring only very slowly in highly competitive industries. Foster for example, in his study of the building industry in which small firms predominate, argued that competition has played at least some part in maintaining the industry's technological backwardness.[16] There are other cases where firms with a major share of the market have been responsible for major innovations. In the UK detergents industry, for example, a study has shown that the two largest firms have been concerned in the development either from the beginning or from a fairly early stage of many of the most important innovations.[17]

Perhaps a more powerful example of this kind is the development of the float glass process in the UK. This was introduced by Pilkington Brothers, which has over ninety per cent of the home market for flat glass. It also has virtually complete control over some specialist glass markets, eg for car windscreens. The Monopolies Commission, which concluded in its study of Pilkington's monopoly position that the company was not operating against the public interest, was impressed with the innovation, which in its words 'speaks for itself'.[18] It is difficult to know, however, in what way it does this, at least as far as the relevance of market structure is concerned, since it cannot be ascertained whether, *under alternative market structures*, innovative activity would have been better, worse or the same. Indeed, market structure may not have been very relevant at all in this particular case : at the time of the innovation Pilkington was a private company; it may be that the close involvement of the Pilkington family in the company led to the necessary substantial development finance becoming available, where as a public company might have been less willing to invest so heavily. The difficulty of making a comparative assessment of the effects of different market structures on innovation, is present in all the case study findings, as we are unable to make 'with/without' comparisons.

More broadly based studies covering industries and innovations may however throw more light on the problem. Mansfield's study (p. 67) is valuable in this respect. As we have seen, the fact that in petroleum and coal the four largest firms did more innovating in proportion to their market share than other firms is not of itself very helpful in determining how important monopoly power is in assisting innovation, because the market shares of the four largest firms differed between the industries chosen. However, using Mansfield's data, it is possible to compare the four-firm concentration ratio with the innovation share-to-market share ratio.

The figures suggest that the more concentrated the industry the smaller is its innovation share-to-market share ratio, although the smallness of the sample must lead to caution in interpreting the data. Statistical studies covering an even wider range of industries are discussed in the section below on 'Concentration ratios and technological change.'

So far, we have considered only the relationship between the market power of individual firms and innovation. What about industries in which *agreements* between firms are in force? Again, the evidence is mixed. For example, members of the Electric Light Manufacturers' Association in the UK agreed, up to the Association's dissolution in 1967, to forego any competitive advantage they might obtain over each other by deciding not to develop any different types of lamp or superior manufacturing methods.[19] One area in which these restrictions were particularly important was in relation to the length of lamp life : an ELMA rule still in force in 1950 restricted the life of a particular type of lamp to a maximum of 1000 hours with fines for excess life and for exceeding a limit on luminous efficiency. Many other examples of this form of agreement could be cited. In the US, for instance, the Radio Corporation of America was formed by three major companies to co-ordinate patent portfolios. Maclaurin has argued that this slowed down the rate of technical progress.[20]

To balance the picture, it is possible to give examples of agreements through which technical progress appears to have been fostered. In 1962 the Permanent Magnet Association appeared before the Restrictive Trade Practices Court in the UK. The association defended its price agreement on the grounds that it fostered technical co-operation and a higher rate of technical progress. The Court accepted the argument, holding that : '. . . the association's technical co-operation had made available to the public more efficient, better designed magnets from a wider number of manufacturers generally more quickly and sometimes at lower prices than would otherwise have been the case . . .' (ie : in the absence of a price agreement).[21] These arguments have also been accepted in other cases (see p. 199).

It is clear from the above discussion that no one particular market structure has an exclusive claim to be the most conducive environment for technological change. For reasons that are discussed below, it is likely that many other factors (eg size of firm in relation to R & D costs) apart from market structure are also likely to affect the pace of innovation and that these factors will vary from industry to industry. This is the clear implication of the statistical investigations discussed in the next section.

4.2.1 *Concentration ratios and technological change*

The discussion so far has been based on specific examples drawn from case study material or data from a very limited number of industries. Although such examples may be useful for illustrative purposes and may form the basis for a more generalised judgement if enough exist, they will often tell us little more than that firm X, which was either a monopolist or a small competitor in a *particular* industry, was also an innovator. More wide-ranging studies are needed if more general conclusions are to be reached. Several economists have explored the relationship between concentration ratios and various measures of technological change for a number of industries. Although concentration ratios may be only a very crude measure of market power, these studies provide a useful, albeit rough, test of the Schumpeterian hypothesis. It should be stressed again, however, that Schumpeter's position would be entirely consistent with a high innovative propensity being related to a threshold level of market power. Furthermore, Schumpeter did not limit his innovation to technical change alone; hence the absence or presence of a close relationship between concentration and, say, R & D expenditures will not provide a direct indication of the relationship between market power and other forms of innovation. However, it might be expected that the different forms of innovation would be affected by the same influences in a similar manner.

The relevant studies divide into two main classes: those using *input* measures of technological change, and those using *output* measures. Of the first kind, one of the earliest studies was by Horowitz.[22] For the twenty-nine American industries he studied, he found a 0.38 rank correlation significant at the five per cent level between industrial concentration, using four firm sales ratios, and research intensities measured in terms of R & D expenditure as a percentage of sales in 1947–48. Hamberg conducted a very similar study for seventeen industry groups in 1958: he found a rank coefficient of 0.36[23] (although this was not significant at the five per cent level). Both writers are careful to stress the limitations of their approach. It is clear that their studies do not provide any evidence for a strong link between industrial concentration and research intensity.

Investigations attempting to relate the *output* of the process of technological change and concentration have had mixed results. Phillips, for instance, has looked at the relationship between twenty *establishment* concentration ratios and technological change as measured by rates of decrease in employment per unit of output for twenty-six American industries.[24] He found a rank correlation coefficient between the 1904 and 1939 establishment concentration ratios and average annual rates of decrease in the indexes of wage-earners per unit of output in 1899–1939 of 0.59 and 0.37 (the first was significant at the five per cent level; the second nearly so). His findings are supported by those of Carter and Williams which were based on British data.[25] Stigler, however, has come up with opposite conclusions. Using the same basic data as that employed by Phillips, he found that productivity increased *faster* in unconcentrated industries than in highly concentrated ones.[26] (It increased even more rapidly in those industries in which concentration *declined* considerably between 1904 and 1935.) It is doubtful whether this difference can be explained solely by the fact that, unlike Phillips, Stigler used *firm* concentration ratios.

The mixed results obtained from these studies suggest that other factors besides monopoly power are at work in influencing the rate of technological change in different industries. The expected profitability of an invention, the speed at which imitation can occur, the number of actual and potential rivals and the magnitude of R & D costs are all likely to have their influence. Another important explanatory factor may be differences in 'technological opportunity' between industries. Several studies have tried to take this last variable into account. Comanor for instance divided his industries into two groups, depending on the possibilities for product differentiation.[27] Phillips used indices purporting to show the strength of the association between the organised sciences and the technologies of the respective industries. Each product group was assigned a code of 1, 2 or 3, depending on 'an obviously subjective evaluation of the extent to which current science permits functional (as contrasted with stylistic) product changes and production differentiation among firms'.[28] Scherer classified his industries to one of four broad product technology classes.[29] In all these studies a positive relationship between concentration and R & D intensity was found, but it was much weaker than when no account was taken of technological opportunity. This result arises because research intensities and concentration are both affected by technological opportunity. The correlation between research intensities and concentration may therefore contain a spurious element.

The causal pattern in the relationship between technological opportunity and concentration has been briefly discussed by Scherer.[30] He suggests that there is a case for arguing that innovation associated with technological opportunities *has led to* concentration, on the grounds that many of the highly concentrated industrial structures in research intensive industries today have resulted in part at least from patent and know-how barriers to entry. Nevertheless, it must be remembered that the results still provide some support for the Schumpeterian hypothesis. Scherer also found that the correlation between research intensities and concentration was strongest in the low-opportunity fields. This may suggest that where the potential profit from an innovation is relatively small, as is likely to be the case in a low-opportunity field (because the costs of R & D will be relatively high and the pay off relatively low), the fewer the number of rival firms who will want to share this profit, the more likely is a firm to commit resources to bring the invention to fruition. On the other hand, where opportunities are high and there is room for everyone, a concentrated market structure will be much less important (indeed, it may have a negative effect on innovation).

4.3 Technological barriers to entry

One important factor that will affect the pace of technological innovation in an industry will be the ease with which new firms, and firms in established industries, can enter. The effect of entry or potential entry may be two-fold. First, it may act as a stimulus to existing firms to innovate in order to maintain their position. However, the reverse may be true: where an innovation can be easily copied, potential entry by outsiders may *reduce* the incentive for established firms to innovate on the grounds that any lead that they may gain as a result will be quickly eroded by innovators. Secondly, new entrants may themselves be an important source of innovation (see p. 65). Numerous studies covering a wide range of industries have shown the often crucial importance of new entrants as a source of innovation. The history of the hovercraft (see part 3) also illustrates this.

Barriers to entry may take several different forms.[31] They may result from the existence of economies of scale in production which require a new entrant to have a sizeable share of the market before it is competitive in cost terms. Strong consumer loyalty to the products of existing producers may form another barrier which only a substantial and costly sales effort by the entrant can overcome. He may also have to sustain financial losses in the first few years of production. Existing producers may control the best sources of supply or the best outlets. For example, in the early days of marine radio, Marconi attempted to fortify its market position (unsuccessfully, as it turned out) by refusing to accept messages at its shore stations, of which it had a far greater number than its competitors, from ships using the equipment of other companies.[32] Absolute capital requirements for efficient production may also be beyond the resources of potential entrants or involve them in obtaining finance at disadvantageous rates. Barriers may also be established by patents and by secret technical know-how. If existing companies possess superior technological knowledge of product processes to which potential competitors have no or limited access, they will *ceteris paribus* have a cost advantage over potential entrants. The firms possessing this knowledge may then attempt to maintain their position by, for example, buying up all the patents in the field, or taking over innovating companies. Thus, patent protection that stems initially from innovative work may (somewhat paradoxically) lead to lethargy in a company's search for new products and processes.

The available evidence on entry barriers is rather sketchy. The most comprehensive study to date is still that by Bain which examined entry barriers in twenty American industries in the early 1950s.[33] He found that barriers arising from consumer loyalty and product differentiation were the most frequent in occurrence (especially as a source of *high* entry barriers); scale economies were of 'slightly lesser importance' and were usually less extreme in their effects, and absolute cost barriers, including patent barriers and ownership of sources of materials but excluding large capital requirements, were 'distinctly less important'. He found that large capital requirements may impose a further entry barrier in a 'considerable proportion of industries'. As Bain points out , however, his study did not cover every industry, and this may limit the generality of his conclusions. For example, his investigations did not embrace chemical, electrical and electronic products, where patents may be an important source of entry barrier. Certainly, in the case of pharmaceutical products, patents appear to have provided manufacturers with strong market positions at least in the short run (see p. 41).

Not all the barriers discussed above will be of a permanent nature. For example, patent rights expire after sixteen years in the UK and licences may be granted before that if they are being unreasonably withheld (see pp. 39–40). If the firm is unable to consolidate its position further, the expiry of a patent may cause a dramatic fall in the market position of a former dominant firm. Seventeen years ago, for example, two American companies, Pfizer and Lederle, were among the top three suppliers to the National Health Service (NHS) in Britain. This position was due to their highly successful products, tetracycline and aureomycin. Since the patents on both products have expired, these two companies have lost their lead and were seventeenth and thirtieth respectively in the NHS suppliers' league table in 1972.[34] Beecham, a British company with a big interest in pharmaceuticals, owes much of its success in this field to its discovery and patenting of just one group of antibiotics: semi-synthetic penicillins. Unless it can widen its range of pharmaceutical products, either internally or through merger, its market position may suffer considerably when the basic patents expire. This fear was undoubtedly a factor in its abortive attempt to take over Glaxo in 1971.[35]

Most of the above discussion on entry barriers has been couched in terms of a (broadly) *given* technology. Even when an outsider tries to enter on the basis of an innovation, these barriers may still prove effective impediments because the innovation may not fundamentally affect the conditions of production, eg the need for certain types of materials, know-how, etc. On the other hand, an innovation may destroy previous existing barriers. Indeed, the existence of the original barrier may act as a *stimulus* to outsiders to produce an innovation which will overcome it.

4.4 Conclusions

We have seen that both the theoretical and the empirical work on market structure and innovation is inconclusive. The pace of innovation is clearly affected by a whole range of influences which will vary from industry to industry. However, it does seem clear that there is little evidence to support a strong interpretation of Schumpeter's hypothesis. One possible interpretation of the data might be that a threshold level of concentration exists: that up to a certain point greater concentration encourages firms to spend on R & D and to innovate, for example by providing greater liquidity and financial incentive and more certainty about the future, but that after this point even great concentration dampens the motivation to innovate. (Stigler's findings (p. 85) are not inconsistent with the idea of a threshold since his unconcentrated industries cover a wide range of concentration levels, the highest being 45.8 per cent.) Again, however, this threshold level, if it exists, is likely to vary from industry to industry. Only when other important effects on the pace of innovation have been isolated will it be possible to test this hypothesis conclusively.

Entry, or potential entry, into an industry is likely to stimulate that industry's innovative activity. Established firms may be protected from such competition by barriers to entry, although in the long run these are likely to become ineffective. They may in fact act as an *inducement* to outsiders to innovate.

NOTES on Chapter 4

1. J. A. Schumpeter *Capitalism Socialism and Democracy* 5th edn London, Allen and Unwin (1952) p. 84
2. S. G. Sturmey *The Economic Development of Radio* London, Duckworth (1958)
3. E. Mason 'Schumpeter on Monopoly and the Large Firm' *Review of Economics and Statistics* (1951)
4. J. A. Schumpeter, op. cit. p. 87
5. F. M. Scherer 'Firm Size, Market Structure and Output of Patented Inventions' *American Economic Review* (1965)
6. J. A. Schumpeter op. cit. p. 83
7. R. M. Cyert and K. D. George 'Competition, Growth and Efficiency' *Economic Journal* (1969)
8. J. W. Markham 'Market Structure, Business Conduct and Innovation' *American Economic Association Papers and Proceedings* (1965)
9. K. Arrow 'Economic Welfare and the Allocation of Resources for Invention' in National Bureau of Economic Research *The Rate and Direction of Inventive Activity* Princeton, Princeton UP (1962) See also W. Fellner 'The Influence of Market Structure on Technological Progress' *Quarterly Journal of Economics* (1951)
10. H. Demsetz 'Information and Efficiency: Another Viewpoint' *Journal of Law and Economics* (1970)
11. B. Yamey 'The Incentive to Invent: A Comment' *Journal of Law and Economics* (1970)
12. R. M. Cyert and J. G. March *A Behavioural Theory of the Firm* Englewood Cliffs, NJ, Prentice Hall (1963)
13. J. Downie *The Competitive Process* London, Duckworth (1958)
14. A. A. Bright and W. R. Maclaurin 'Economic Factors Influencing the Development and Introduction of the Fluorescent Lamp' *Journal of Political Economy* (1943)
15. R. Miller and D. Sawers *The Technical Development of Modern Aviation* London, Routledge (1968) p. 265
16. C. Foster 'Competition and Organisation in Building' *Journal of Industrial Economics* (1964)
17. W. J. Corlett *The Economic Development of Detergents* London, Duckworth (1958) p. 202
18. *A Report on the Supply of Flat Glass* (Monopolies Commission Report

HC 83) London, HMSO (1968) p. 73
19. *Second Report on the Supply of Electric Lamps, Parts I and II* (Monopolies Commission Report HC 4) London HMSO (1968)
20. W. R. Maclaurin, *Invention and Innovation in the Radio Industry* New York, Macmillan (1949) p. 254. Quoted in F. M. Scherer *Industrial Market Structure and Economic Performance* New York, Rand McNally (1971) p. 371
21. *Reports of Restrictive Practices Cases* vol. 3 p. 122
22. I. Horowitz 'Firm Size and Research Activity' *Southern Economic Journal* (1961/62)
23. D. Hamberg 'Size of Firm, Oligopoly and Research: the Evidence' *Canadian Journal of Economics and Political Science* (1964)
24. A. Phillips 'Concentration, Scale and Technological Change in Selected Manufacturing Industries, 1899–1939' *Journal of Industrial Economics* (1956). Phillips's sample unfortunately excludes a number of important industries.
25. C. F. Carter and B. R. Williams, *Industry and Technical Progress* Oxford, Oxford UP (1957)
26. G. Stigler 'Industrial Organisation and Economic Progress' in L. D. White (ed.) *The State of the Social Sciences* Chicago, Chicago UP (1956)
27. W. Comanor 'Market Structure, Product Differentiation and Industrial Research' *Quarterly Journal of Economics* (1967)
28. A. Phillips 'Patents, Potential Competition and Technical Progress' *American Economic Review: Papers and Proceedings* (1966)
29. F. Scherer 'Market Structure and the Employment of Scientists and Engineers' *American Economic Review* (1967)
30. ibid. pp. 374–5
31. J. Bain *Industrial Organisation* 2nd edn, New York, Wiley (1968) chapter 8
32. S. G. Sturmey *The Economic Development of Radio* London, Duckworth (1958)
33. J. Bain op. cit.
34. *The Times* (14 January 1972)
35. *Beecham Group Ltd, and Glaxo Group Ltd; The Boots Company Ltd, and Glaxo Group Ltd: A Report of the Proposed Mergers* (Monopolies Commission Report, HC 341) London, HMSO (1972)

5 Scientists and Engineers: their Supply and Demand

In previous chapters, the process of technological change has been considered primarily in an organisational context with attention being concentrated on issues such as the effect of a firm's absolute or market size on its inventive and innovative effort. In this chapter the focus switches to the personnel involved. The discussion divides into three parts. In the first section the characteristics of the market for qualified scientists and engineers (QSEs) as a whole is examined, and particular attention is paid to the concept of shortages in this market. The second section deals with the mobility of scientists and engineers between companies and institutions. The final section considers the economic issues involved in the *international* movement of QSEs.

5.1 The market for QSEs

Before considering in detail the market for QSEs, it is appropriate to look at some problems of definition.

5.1.1 *Defining a QSE*

Terms like 'scientist' or 'engineer' are not of course unambiguous descriptions of particular types of persons. Usually the term 'qualified scientist or engineer' used in official surveys is applied to a person with a minimum formal qualification, normally at degree level or equivalent. This creates several problems. Firstly, it is clear that QSEs will vary widely in their skills and abilities. Comparisons between industries, and indeed firms, may therefore be difficult if as is likely, certain industries and firms tend on the whole to attract better QSEs than others. Productivity in research also tends to vary with age : where industries have different age structures among their QSEs, straight comparisons of manpower deployment may give misleading impressions about the extent of the inputs. It may be possible partly to overcome these problems by a weighting system, but it is by no means certain that a large number of mediocre QSEs would *ever* achieve the results of one 'brilliant' researcher.

The second problem arises because the usual definition of a QSE is based on paper qualifications. Yet this practice excludes all those people who are performing scientific and engineering activities but who do not have the requisite qualifications. It is difficult to estimate the extent of this problem. Layard *et al.* have shown however that nearly sixty per cent of those employed in technological jobs in the electrical industry have no degree or professional qualification.[1] Many of these jobs would probably be considered suitable for QSEs. The other side of the coin is that the abilities of some QSEs are under-utilised, so that they are doing jobs for which they do not need their qualifications; or they may be in jobs for which their scientific or engineering training is only necessary for *part* of their work. In the absence of better data however we are forced to use those currently available from official sources.

5.1.2 *'Shortages' in the QSE market*

In the postwar period in Britain a large number of official and unofficial reports have been published which in different ways have been concerned with either the demand for or the supply of QSEs. Topics such as the lack of university places in science and engineering, the swings away from these subjects in the sixth form and the emigration of QSEs have all received thorough airings.[2] Throughout the 1960s particularly the question of shortages of scientists and engineers – particularly the latter – was continually cropping up. In the early 1970s however discussion of shortages became somewhat less marked as graduate unemployment rose.

How can a shortage in a particular skill be identified? In a general sense, of course, a shortage exists wherever a stipulated aim is not met, whether that aim is or is not economically justified. In more economic terms however a shortage is usually considered as occurring when at the existing price (salary) more is demanded than is supplied. This may be caused by a change in supply, resulting for example from a 'brain drain', or by a shift in demand following, say, an increase in government-sponsored R & D. Or it may be caused by a combination of both supply and demand factors. This would tend to show itself initially in an expanding number of vacancies at the existing salary level.

This 'shortage' can be eliminated by raising salaries to their new equilibrium level. This readjustment process takes time, for several reasons.[3] Firstly, a firm may not immediately perceive the true supply–demand situation or realise that vacancies will be filled only if salaries are raised. Obviously the greater the shortage at existing salary levels the more quickly the firm is likely to become aware of the disequilibrium situation. Secondly, the firm will have to determine whether it would be in its economic interests to raise the salaries offered. Thirdly, proposals for changing salaries are not normally approved immediately: they usually have to go through established administrative channels. Fourthly, for a time it may be that existing employees can be paid lower salaries than a new entrant, so that in a given firm a *range* of salaries exists. In time of course it is likely that these differences will disappear, as information filters through to existing employees. The position is further complicated by the fact that market forces are continually changing: a shortage develops, but as adjustments to this shortage are being made a further change in underlying market conditions occurs.

However, it might be expected that *eventually* a shortage of QSEs relative to other occupational groups would be expressed in their salaries rising more quickly. (It should be noted that our concern here is with *relative* shortages; the salaries of all occupational groups may be rising). This hypothesis must be qualified in two respects. Firstly, because of institutional factors (eg incomes policies), salaries may *not* respond to demand and/or supply changes, although this qualification is unlikely to apply with very great force in the UK or US context in the long run. Secondly, salaries are only one, albeit an important, aspect of job remuneration and attractiveness. Other influences such as working conditions, promotion prospects etc can carry some weight and a firm may attempt to increase these other attractions as well as salaries to fill its vacancies.

Blank and Stigler have used the relative behaviour of salaries as a basis for establishing whether a shortage existed for engineers in the United States over the period 1929–56. They found that :

> Since 1929 engineering salaries have declined substantially relative to earnings of all wage earners and relative to incomes of independent professional practitioners. . . . After the outbreak of the Korean war there was a minor increase in the relative salaries of engineers, but this was hardly more than a minor cross current in a tide.[4]

Having satisfied themselves that the market was operating reasonably well in a technical sense, Blank and Stigler argued that their findings showed no shortage of engineers over the period. Arrow and Capron have however challenged the conclusions Blank and Stigler reach for the post-Korean war period, on the grounds that 'there may be a considerable lag in the adjustment of salaries in response to changes in demand'.[5] Arrow and Capron's argument is that, although demand for engineers had risen rapidly and although elasticity of supply is low for at least short periods, this was not reflected in the data because in their view the 'speed of reaction' in the engineer–scientist market is *especially* low. This argument is based on the propositions that long-term contractual arrangements between employees and employers mean that salaries are renegotiated only infrequently; that the engineer–scientist is often unsure of the market to which he belongs and therefore does not know whether higher salaries elsewhere should be taken as a signal to *him* to press for more; and that the existence of a large buyer – government – may slow down the rate of salary increases.

Arrow and Capron's arguments are open to challenge especially as it is by no means certain that they apply *particularly* forcibly to engineers and scientists; but they do support the basic hypothesis that a shortage will express itself in salary rises, albeit only after a time lag. Nowhere do they state however *how long* this time lag is likely to be.

In the UK attempts have been made to see whether a shortage of engineers has existed in recent years by using the Blank–Stigler approach. The latest study, which looks not only at relative salary movements, but also at changes in activity rates, vacancies and stocks, concludes that 'it is hard to make any case for a general shortage of engineers in the 1960s',[6] although there have been occasional changes in relative earnings. The present value of the differentials between the salaries of graduate and non-graduate engineers also seems to have declined at least for 1967/68–1970/71, suggesting that graduate engineers have been in more plentiful supply than their non-graduate counterparts. This narrowing seems to be only one aspect of an overall deterioration in the relative position of graduate manpower as a whole in the later 1960s in the UK.[7]

5.1.3 Substitution between skills

The term QSE embraces a whole range of possible skills and qualifications. Although these skills are often very different it is likely that, when a shortage in one particular type occurs, some substitution between skills will be made. More commonly, less qualified manpower may also be utilised (a non-QSE chemist may be a closer substitute than, say, a QSE engineer, for a QSE chemist). Both forms of substitution almost inevitably involve additional costs. For example, the productivity of a 'technician' may be much less than that of a QSE : in a study of twenty-seven large US chemical and pharmaceutical companies it has been shown that increasing the non-professional-to-professional ratio in R & D has little effect on the outcome of most research projects.[8] Furthermore, outside of R & D less qualified manpower may have more difficulty in ensuring the acceptance of R & D results in production.

Peck has examined the question of substitution of skills and has shown that, although the total technical employment (QSEs and technicians) as a proportion of the total labour force tends to be similar in both the US and the UK, the latter has 4.7 technicians per QSE and the US, 0.62.[9] This difference may be partly due to differences in definitions between the two countries. Furthermore, it may reflect not a market shortage of QSEs in the UK in the sense described at the beginning of the chapter, but rather a demand that is 'irrationally low'.[10]

In the later part of the 1960s there is in fact evidence in the UK to show that substitution was occurring in the opposite direction. In both R & D and in non-R & D activities, a greater percentage of technicians' jobs were being done by QSEs. Between 1965 and 1968 the percentage of all QSEs in R & D manufacturing industry who were working as technicians rose from 13 to 22 per cent. Outside R & D, the percentage rose from 15 to 25 per cent.[11]

5.2 The mobility of scientists and engineers

So far discussion has focused on the characteristics of the market for QSEs as a whole. In this section the operation of this market at the micro level is discussed. The movement of technical personnel between institutions – a process that has been called 'technology on the hoof' – is obviously one way in which demand for QSEs by particular organisations may be satisfied. The discussion will not be restricted to QSEs as defined in the official surveys: the term 'technical personnel' is drawn much more widely and refers to anyone who has a detailed knowledge of a particular production technique or techniques. They may or may not be working in R & D.

The implications of such mobility for industrial competition may be considerable: while competition is traditionally viewed as taking place between 'firms' in some abstract sense, one of its crucial elements may revolve around a highly complex and inter-connected movement of personnel between competitors. This is clearly true of managerial personnel – the growth of 'headhunting' activities and organisations in recent years provides some measure of its importance in this sphere – but it may be less obvious that the movement of technical personnel between firms both inside and outside the R & D function may also have important repercussions on the competitive structure in both the short and long runs. This implies that some scientists and engineers are very difficult to replace and that grouping all of them under a general heading of 'QSE' may disguise the very specific nature of their skills. This may be particularly misleading in the context of a very young technology in which only very few people have been working.

Mobility of technical personnel is also important for another reason: it often plays an important role in the diffusion of an innovation (see chapter 7). New or existing firms who wish to produce a new product or use a new process require scientific and technical manpower with knowledge of the innovation. Rather than 'grow their own', they may recruit employees of other companies already using it.

This section looks at some of the basic issues involved in 'technology on the hoof' where either the recipient or donor is a private firm. *Horizontal* technology transfer (ie between organisations who are operating at, or concerned with, the *same* stage of production) rather than vertical transfer (ie between organisations concerned with different stages) is considered here.

There are numerous ways by which technology can be transferred from one organisation to another. Firstly, the tangible product of one organisation will itself transmit some of the know-how used in its production. How effective such a transfer is depends on the capabilities of the organisation receiving it in relation to the complexity of the product, and the extent to which the know-how used in the product can be concealed. Secondly, technology may be transferred by the written word, through journals, papers, patent specifications, etc. The efficiency of this transfer mechanism will depend on the abilities of the writer and the reader. Some information may be withheld because of commercial or national secrecy or, if it is not withheld, may not be usable in the form in which it is presented. We have already seen that patents often only provide a partial disclosure of information (see p. 39).

The third way is through temporary contact, visual and/or verbal, between personnel from different organisations, through conferences, seminars, telephone calls, etc. Such contact may vary in length from short meetings to extended 'secondments'.* Firms often work in close technical collaboration, even though this stops short of integration by ownership.[12]

The fourth method of transfer is the one that concerns us here. This involves the transfer of personnel from one organisation to another which at the time of transfer is intended to be permanent. For the receiving institution this method is likely to be superior to the third method of transfer outlined above. It inevitably involves movement of experience and know-how that would otherwise remain in the 'donor' institution. This follows directly from the fact that technical know-how has become embodied in the person concerned, and that legal or other limitations on the use of such

---

* Our discussion here is restricted to technology. However, contact between workers in basic research may also play an important part in stimulating ideas. For example in the development of matrix isolation, 'progress was made through direct contact among the many scientists involved resulting in the exchange of information concerning techniques and negative results'. *Technology in Retrospect and Critical Events in Science* Washington, Illinois Institute of Technology Research Institute, Washington, National Science Foundation (1968) p. 17.

know-how in the donor institution are unlikely to be fully effective. Some indication of the importance of technology on the hoof is given by Langrish et al. in their study of 51 innovations that received the Queen's Award to Industry in the late 1960s in Britain.[13] They analysed the sources of ideas leading up to the innovations, and found that of the 158 important ideas they were able to identify, 102 came from outside the firm. This fact in itself gives some indication of the importance of the transfer mechanism generally, a finding that is confirmed by other studies. But more important for our purposes is their analysis of the method of transfer of these 102 ideas. This is shown in table 5.1. It is clear that the most important single route for such a transfer was through technology on the hoof.

Table 5.1
Method of transfer of 102 important ideas from outside the award-winning firm

Transfer via person joining the firm	$20\frac{1}{2}$*
Common knowledge: via industrial experience	15
via education	9
Commercial agreement (incl. take-over and sale of know-how)	$10\frac{1}{2}$*
Literature (technical, scientific and patent)	$9\frac{1}{2}$*
Personal contact in UK	$8\frac{1}{2}$*
Collaboration: with supplier	7
with customer	5
Visit overseas	$6\frac{1}{2}$*
Passed on by government organisation	6
Conference in UK	$2\frac{1}{2}$*
Consultancy	2
	102

* The '$\frac{1}{2}$s' indicate that in the case of some ideas, the sources mentioned were not mutually exclusive.
Source: J. Langrish et al. *Wealth from Knowledge* London, Macmillan (1972) table 7

In some instances organisations themselves may undergo change, through takeover and/or merger, to facilitate transfer. For example, in its analysis of the (abortive) attempt by Beecham to take over Glaxo in 1971, the Monopolies Commission found that the former had 'a strong incentive to remedy its potential weakness [ie its narrow research base] by broadening the scope of its research'.[14] This would have been achieved by Beecham absorbing Glaxo's research staff.

The 'Bolton' Committee of Inquiry on Small Firms recognised that technology on the hoof might be an important factor in innovation:

> Possibly the majority of innovations . . . arise from a little understood process of movement of people: through small firms being acquired by larger firms or licensing larger firms or acquiring licences from them; by staff leaving one company and joining another and also through the so-called spin-off process where an individual leaves a large company, university or laboratory (sometimes because he cannot get his ideas implemented) to set up his own business, perhaps to succeed spectacularly, perhaps to fail and possibly in either event, to sell out to a larger firm.[15]

It may of course be argued that the need for technology transfer reflects the division of labour in the R & D–invention–innovation process between different types of production 'units'. It is probably true for example that small firms and independent individuals *generate* a disproportionate share of all new ideas and inventions which the large firm then takes up. This in turn will require transfer of technology, often on the hoof. In other cases, innovation is excluded by definition from an organisation's activities. A university or research association *must* transfer its technology by some means if innovation is to occur.

5.2.1 *Donor and recipient institutions*

We are concerned here only with the way in which an individual transfers from one institution to another. Intra-organisational transfers are ignored, although their importance, for example in obtaining acceptance of R & D output in production or marketing departments, is readily acknowledged. The basic flows of technology on the hoof discussed here are given below (in some cases the donor or donee institution will be synonymous with the individual himself):

1) public to private institution;
2) private to public;
3) private to private.

'Public institution' covers university and government laboratories and departments and non-profit-making research organisations. 'Private institutions' can range from the small, one-man concern to the large corporation. The National Research Development Corporation (and its subsidiary companies) in the UK presents classification difficulties, but because it is heavily dependent on government funds (see p. 146), it is classified here as a public institution.

Under category 1) above, there are well-documented cases of persons leaving public institutions either to set up on their own or to join another company. Roberts for example has examined the new companies founded by ex-employees of several laboratories and academic departments of the Massachusetts Institute of Technology, of a government laboratory and of a non-profit-making organisation in the late 1960s.[16] Some idea of the numbers involved is given in table 5.2.

Table 5.2
Sources of new technical enterprises

Sources of new enterprises	New companies identified
MIT laboratories	
Electronic Systems Laboratory	11
Instrumentation Laboratory	30
Lincoln Laboratory	50
Research Laboratory for Electronics	14
MIT academic departments	
Aeronautics and Astronautics	18
Electrical Engineering	15
Mechanical Engineering	10
Metallurgy	8
Government laboratory	
Air Force Cambridge Research Laboratory	16
Not-for-profit organisation	
MITRE Corporation	5
Industrial electronic systems contractor	39

Source: E. B. Roberts 'A basic study of innovators; how to keep and capitalize on their talents' *Research Management* (1968)

Most of the firms studied by Roberts have so far proved successful. Other examples could be quoted of such spin-offs, often arising from a combination of the non-innovating nature of public institutions and the reluctance of existing private companies to take up an idea or invention.

An employee of a public institution may transfer directly to an existing organisation. Langrish *et al.* quote the example, among others, 'of Sims and Briggs who were recruited to join Davy and

United to prepare for commercial exploitation the system for automatic control of steel strip thickness on which they had already worked in the British Iron and Steel RA'.[17] In the hovercraft industry in this country, several employees of Hovercraft Development Ltd, a subsidiary of NRDC, have set up their own companies or joined established companies (see part 3).

Some reverse movement (category 2)) also takes place. Numerous examples exist of academics and government scientists and technologists who have been appointed to their present posts after experience in industry.

Category 3), which is probably the most interesting from an economic point of view, may involve the setting up of new companies by ex-employees of other companies. For example, Scherer points out that thousands of research-based firms in the United States have been set up by ex-employees of large companies, such as Sperry Rand, IBM and Western Electric.[18] Roberts's study (above) showed that thirty-nine companies had been set up by forty-four ex-employees of one large Greater Boston electronics company alone.

From Bell laboratories in the United States, fifteen firms have been established in less than twenty years, as table 5.3 shows.

In some cases, transfer of staff will take place between firms that are *already* competing with each other in a particular field. In others a firm may want to expand its range of activities and may seek to recruit from firms already engaged in the relevant area. For example, when Vosper Thornycroft entered hovercraft production it obtained its chief designer from the British Hovercraft Corporation, the largest company in this field. In the development of the Tube and Tank process in the oil refining industry in the United States, 'Jersey Standard relied for technical skill almost entirely upon the employees whom it hired away from Indiana Standard – Clark, Loomis, Carringer, at that time the most inventive and least contented men in Indiana's organisation'.[19] Another study that has highlighted the mobility of technical personnel between firms in the US is that by Tilton on the semiconductor industry. Interestingly, he found however that mobility in Europe was appreciably less than in the US. Tilton suggests that among the reasons for this relatively lower mobility is a difference in business ethics and the somewhat smaller number of semiconductor firms in Europe.[20]

SCIENTISTS AND ENGINEERS : THEIR SUPPLY AND DEMAND 103

Table 5.3
Semiconductor firms descending from Bell Laboratories, 1952–67

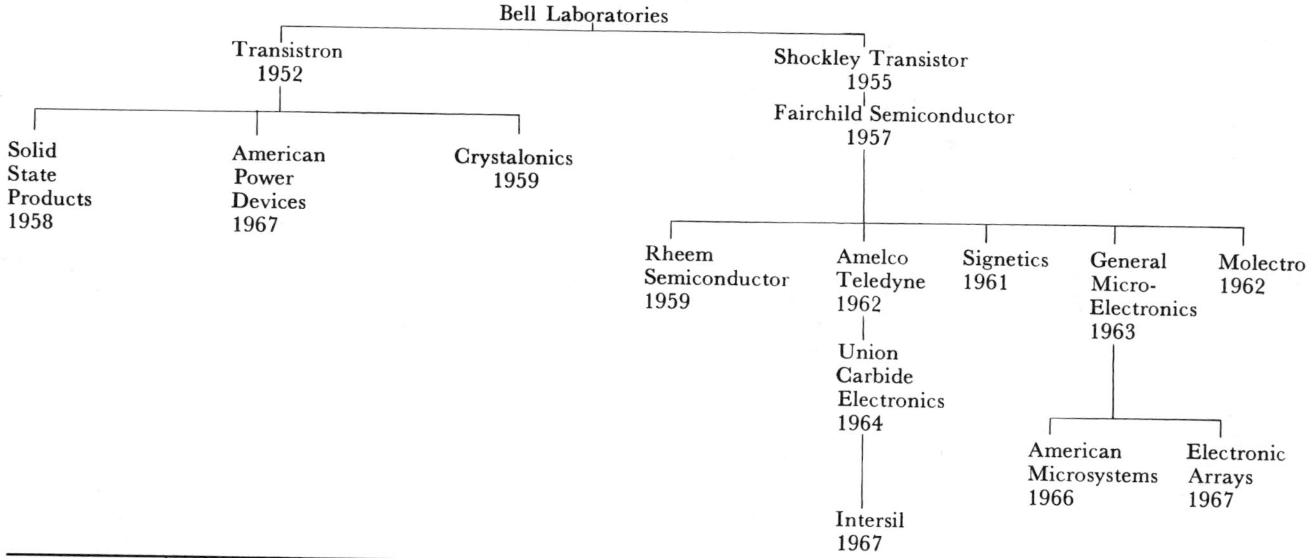

Source: D. C. Hoefler 'Semiconductor Family Tree' *Electronic News*, (July 1968) and other trade sources, quoted in J. E. Tilton *International Diffusion of Technology: The Case of Semiconductors* Washington, Brookings Institution (1971) figure 4.1 (© 1971 Brooking Institution)

The reasons why technical personnel are prepared to change firms are of course many and varied. Higher salaries may play an important part but other factors, ie greater authority and power and the opportunity to develop hitherto frustrated ideas, are also likely to play their role (see part 3).

5.2.2 Legal restrictions on the transfer of technology on the hoof

Three important legal restrictions may be imposed on the transfer of technology on the hoof. Firstly, the requirements of national secrecy may affect the degree to which know-how can be utilised elsewhere. This may be particularly important in type 1) transfers, but it may also be important in others. This issue is not discussed here. Secondly, if company A holds unexpired patents on an invention, then there may be clear restrictions on its use by company B, even though technical personnel may have transferred from A to B. The common law position is that inventions produced in the normal course of employment belong to the employer.[21]*

How far the refusal to license a patent restricts the transfer of technology on the hoof will depend on numerous factors. The way in which the patent specification has been drawn up, and the ability of a competitor to 'design round' are two factors of obvious importance here. The willingness of a patent holder to go to court to enforce his rights will be another factor. If there are any doubts at all about the patent then legal enforcement may not be pursued. If on the other hand a company unquestionably holds the basic patents for a process then it may be very difficult for another company to circumvent them, although it is certainly not unknown.[22] Patents do not of course last for ever; they have a minimum life of sixteen years, with a possible extension in the UK to twenty-six years. There is then no restriction on the use of the invention. Where a licence is provided transfer will be much more easily accomplished.

Thirdly, the common law provides a company with certain rights over its own know-how, 'technical knowledge of industrial significance which has been built up in one organisation and is not in the public domain'.[23] (Many patent agreements also include

---

* It is not of course always clear whether an invention has been produced in the normal course of employment. See reference in n.21 for a fuller discussion.

terms on the provision of know-how.) Employees cannot divulge know-how of the trade secret or proprietary type without permission. Such information belongs to their employers. This restriction applies when employees move to another company. It appears however that once such information is divulged, there is no right of claim against those using the know-how, even though there may be a right of action by a company against the individual ex-employee who discloses it. Ownership of know-how differs from patent rights primarily on the grounds that *once the know-how has reached the public domain* there is no restriction on its use.

Although these common law rights exist, considerable problems arise in any attempt at enforcing them. It is often impossible to separate that know-how that is protected by common law and that experience skill and knowledge that is acquired by an employee in the normal course of his employment. Even if a clear distinction is possible it will usually be difficult for a company to prove that an ex-employee working in another company has disclosed the confidential information. And even then the damage has been done, since competitors are free to use this knowledge. A company may attempt to stop technology from transferring to a competitor by putting a term in an employee's contract of employment which restrains the employee from working for a competitor if he leaves. Normally however such a term would be void as being in 'restraint of trade'. However there is nothing to stop an employer from introducing factors that are aimed at reducing mobility : pension rights and salary scales, for example, may all be geared to length of service.

5.2.3 *Some economic issues*

The discussion in this section will be limited to the *firms* involved.

The donor firm
The way in which, say, an engineer acquires knowledge of a particular branch of technology is in many respects akin to a training process. As this knowledge is absorbed, so that person's productivity is likely to rise as he becomes more able to contribute to the particular technology involved, assuming that such familiarity does not stifle creativity. Productivity in this context is extremely difficult to define, but in broad terms it may be said to be reflected in the

employee's contribution to the profitable development of his employer's products.

Like all training processes, it may involve costs apart from salaries – for example of going down 'blind alleys' – and it is unlikely that the engineer's productivity initially at least will be able to compensate for them. Becker has defined two extreme 'poles' of training : completely *specific* training is 'training that has no effect on the productivity of trainees that would be useful in other firms', while completely *general* training 'increases the marginal productivity of trainees by exactly the same amount in the firms providing the training as in other firms'.[24] Clearly, most types of training will be found between these two extremes, but training with different biases will, according to Becker, give rise under competitive conditions to different types of firm behaviour in relation to its financing.

In the context of this chapter however it must be remembered that, even though at a given time only firm A may be using a particular process or product or have secret know-how, firm B may wish *to enter the field*, and to acquire the necessary know-how and/or personnel. In this case the distinction between general and specific training will hinge primarily on complex questions of the legal interpretation of property rights in certain types of knowledge and of patent enforcement.

If a firm loses a key worker to a competitor, then it may still achieve a return on its 'training' investment from the monopoly profits it can earn on inventions, or from proceeds from agreements on patents and the transfer of know-how. This is of course the economic basis of the patent system (p. 42). Nevertheless the problem of defining and enforcing property rights in the knowledge a transferee has is considerable, especially in non-patented areas of knowledge. Furthermore, the departing employee may not be immediately replaceable and the donor might find that its innovative ability is thereby affected. On the other hand there may be some benefits from mobility. Tilton for example has argued that Bell Laboratories receives 'useful information through the friendly, informal relations it maintains with former employees scattered throughout the industry'.[25]

The recipient firm
The firm that wishes to enter a particular field of technology but does not have the requisite expertise will have to assess the costs

involved in acquiring that expertise internally, and set them against the costs of bidding away an experienced worker from another company. Costs of acquiring a technology internally will of course vary according to the particular field involved. This learning process may be telescoped if someone with the necessary expertise can be hired through the market. Costs of absorbing the new worker however remain. Furthermore, it may also be true that in some contexts hiring *one* worker is not sufficient, if that worker has previously been working as a member of a team. And as we have seen, attempting to bid away a worker may only lead to a general rise in all salaries with no net gain in employment by the firm that initially wanted additional labour.

It must be remembered again that, although a firm cannot usually stop its employees from being employed by a competitor, it may use patent and common law rights to restrict its competitor's activities. This will inevitably affect the relative attractiveness of the alternatives open to the potential recipient firm. It will not however affect the situation where the transferee is hired for his *potential* productivity rather than for his existing technological knowledge.

Despite the possible savings obtainable from hiring a competitor's employee, this poaching has often been of a limited magnitude. This may be due to a fear that the donor firm will retaliate. In the US oil industry, for example :

> It is surprising that the oil companies did not go to greater lengths in hiring extremely capable employees from their competitors. By inducing a genius to change his job a company could, with the expenditure of a few thousand dollars, eventually save many millions; . . . Possibly the companies did not wish to disturb the existing salary pattern. This is not a very persuasive explanation, however, for potential innovators seem less concerned with increases in salaries, than with working in an organisation that they find hospitable. Possibly the company managers felt that they would lose as many able employees as they could gain if their competitors became predatory too.[26]

Arrow and Capron have argued that the desire to avoid competitive bidding – especially if the total supply of engineers–scientists is unlikely to change much in response to higher prices – will tend to stop any one firm in an industry dominated by a few large ones from increasing its salaries to attract scientists and engineers.[27] This may lead to 'no raiding' agreements being made between otherwise competing firms, although these agreements are unlikely to last indefinitely because of competition from smaller firms.

5.3 International movement of scientists and engineers

The previous section considered the movement of scientists and engineers between firms and institutions and its implications for competition, without reference to national boundaries. However, there are obvious differences in policy implications between intra- and international movements, even though much of the earlier analysis may still apply. This section looks briefly at some of the economic implications of the international movement of qualified manpower.

Scientists and engineers tend to have a higher degree of international mobility than many other occupational groups. Their skills usually have wide applicability and are in demand throughout the world. Where salary differentials and differences in working conditions, prospects, etc exist between countries that are sufficient to overcome any opposition to movement on other grounds, then there will tend to be net migration to the more attractive country, with the donor country experiencing a 'brain drain'. Britain found itself in this position in the first part of the 1960s. An official investigation into the causes and effects of this movement found that the net loss had increased rapidly over the period 1961–66, and that most emigrants had gone to the US. The outward flow was particularly marked among engineers.[28] The causes of such movement are complex. As the report of the investigation stated, salary differentials were clearly important – in the US a young graduate or PhD could earn two and a half to three times that offered in the UK in real terms – but other factors, such as working and living conditions and promotion prospects, were also clearly relevant.

What are the economic implications of such movement for the donor country? Firstly, there are the short-term effects. If a scientist or engineer leaves he may create a gap that can be filled only by someone of inferior ability, or perhaps cannot be filled at all until a suitable replacement can be trained, which may take several years. Teamwork may be disrupted if an important member leaves and it may be very difficult to recreate the earlier cohesiveness. In this case the efficiency of the remaining team member will be decreased. However the problems involved will be somewhat reduced if, as is likely, the company concerned knows well in advance that an employee is going to resign and is able to plan accordingly. Most companies in any case plan for some measure of turnover. The disruption effects will clearly also be dependent on the importance of the individual who leaves, and on the 'reshuffling' of the remaining employees that then becomes necessary.

In the longer term the effects of migration are much more complex and have been the subject of controversy.[29] At first sight it seems reasonable to assume that if the QSE is paid the value of his marginal product, when he emigrates the *per capita* income of those remaining behind in his country of origin would remain unaffected. When emigration occurs output may fall, but so does the income payable. The net effect is therefore zero.

However this view must be qualified in several respects. Firstly, the emigrant may not have been paid the full value of his worth to society, ie externalities may be present. But as Grubel and Scott have pointed out, if these external benefits are linked more to the jobs involved, rather than to a particular individual, then emigration may involve only the short-term adjustment costs incurred until a replacement, who will also provide external benefits, is found.[30] (The same arguments apply to external diseconomies.)

Secondly, where the scientist or engineer pays taxes which are then, in part or in whole, redistributed to others or used to provide services that he does not receive, then his emigration will cause some loss to those remaining behind. For example the emigrant's tax payments may have been used in part to finance pensions for others. It must be remembered however that the emigrant also relieves the donor country of caring for *him* in old age (unless of course he returns). Even in the situation where his tax payments are fully returned in kind or in cash, through the social services, his move to another country may not mean that the services can be reduced proportionately because of indivisibilities in their provision.

Thirdly the movement of knowledge across national boundaries may mean that the donor country becomes *better* off. For example if scientists working on basic research in country A move to country B to do similar work, country A will still be able to obtain the results of such research through learned journals etc without paying the costs involved. In this sense a country will benefit from a brain drain. Furthermore if, as Grubel and Scott point out, conditions for carrying out such research are better in the receiving country then the gain will be even greater.

There are at least three possible qualifications to this argument. Firstly, the maintenance of a basic research effort may be necessary for all-round scientific awareness and for maintaining the ability to appreciate the implications of new developments. Thus a country receiving results from another may still have to maintain its own basic research capability. Secondly, transmission of knowledge inevitably takes time; this may give the country actually carrying out the research a valuable lead time. Usually however, in the case of basic research at least, this benefit is likely to be marginal. Thirdly, even with fundamental work there are likely to be different national biases in objectives. Thus work undertaken in one country may not be very relevant to the needs of another. In applied research and development the international transmission of knowledge is much slower and can be controlled more rigorously, although not completely, by industrial secrecy and patent rights (see p. 172). Migration in this field may therefore give rise to longer-term costs as well as to short-term adjustment costs. However, it may still be a rational economic policy for a country to 'buy in' technology rather than create it itself (see p. 22). Some economists have also argued that a donor country is not able to recoup the human investment made in the education of the emigrant: that the receiving country obtains the returns from that investment without paying for it. There is some force in this argument, although it must be realised that the emigrant also removes the burden of the education of *his children* from the donor country.

It can be seen that net effect of emigration on the donor country as a whole is very difficult to determine. In some cases it may be very small.

For an individual donor company, the effects of emigration could be analysed in a similar way. They will depend crucially on the part played by the individual in the company and on the possibilities for obtaining a replacement.

5.4 Conclusions

The distribution and supply of and demand for scientists and engineers is obviously a very important element in the process of technological change. Shortages of QSEs are therefore of clear consequence. Market imperfections and other factors make the correct diagnosis of a market shortage difficult; if, however, relative salary increases are taken as a guide, then contrary to what is normally believed, it does not seem as if there has been any shortage of engineers (over which the primary concern has been) in the 1960s at least in the UK.

At the micro level, the mobility of technical personnel between institutions often seems to play an important part in the innovation process. Although there are however some legal limits on the extent to which technology may be transferred 'on the hoof', these are unlikely to be fully effective. Economic factors may also be at work in reducing inter-firm mobility : firms may be reluctant to bid 'key' personnel away from competitors for fear of retaliation.

On the international front, the net effects of migration on the donor country cannot be determined from *a priori* reasoning, but will depend on the particular job circumstances of the emigrants.

NOTES on Chapter 5

1. P. R. G. Layard *et al.* *Qualified Manpower and Economic Performance* London, Allen Lane (1971), quoted in G. C. G. Wilkinson and J. D. Mace 'Shortage of Supply of Engineers: A Review of Recent Evidence' *British Journal of Industrial Relations* (1973)
2. See for example *Scientific Manpower* Cmd. 6824; *The Brain Drain* Cmnd. 3417; *Enquiry into the Flow of Candidates in Science and Technology into Higher Education* Cmnd. 3541. For a more detailed analysis of some of these issues see M. Sanderson *The Universities and British Industry* London, Routledge and Kegan Paul (1972) chaps. 12 and 13
3. The following discussion draws heavily on K. J. Arrow and W. M. Capron 'Shortages and Salaries: The Engineer–Scientist Case in the United States' *Quarterly Journal of Economics* (1959)
4. D. Blank and G. Stigler *The Demand and Supply of Scientific Personnel* New York, National Bureau of Economic Research (1957) p. 28
5. K. Arrow and W. M. Capron op. cit.
6. G. C. G. Wilkinson and J. D. Mace op. cit.

7. E. G. Whybrew 'Trends in the Labour Market for Highly Qualified Manpower' in H. Greenaway and G. Williams (eds) *Patterns of Change in Graduate Employment* London, Society for Research into Higher Education (1973)
8. Quoted in M. Peck 'Science and Technology' in R. Caves *et al. Britain's Economic Prospects* London, Allen and Unwin (1968) p. 458
9. *ibid.* pp. 456–7
10. *ibid.* p. 460
11. *Persons with Qualifications in Engineering Technology and Science 1958 to 1968* London, HMSO (1971) table S17
12. G. B. Richardson 'The Organisation of Industry' *Economic Journal* (1972)
13. J. Langrish *et al. Wealth from Knowledge* London, Macmillan (1972)
14. *Beecham Group Ltd and Glaxo Group Ltd; The Boots Company Ltd and Glaxo Group Ltd A Report on the Proposed Mergers* (Monopolies Commission Report, HC 341) London, HMSO (1972)
15. *Report of the Committee of Inquiry on Small Firms* (Cmnd. 4811) London, HMSO (1971) p. 53
16. E. B. Roberts 'A Basic Study of Innovators – How to Keep and Capitalize on their Talents' *Research Management* (1968)
17. J. Langrish *et al.* op. cit. p. 45
18. F. M. Scherer *Industrial Market Structure and Economic Performance* Chicago, Rand McNally (1970) p. 354
19. J. L. Enos *Petroleum Progress and Profits: A History of Process Innovation* Cambridge, Mass., MIT Press (1962) p. 231
20. J. E. Tilton *The International Diffusion of Technology: The Case of Semi Conductors* Washington, Brookings Institution (1971) p. 119
21. For a discussion on this issue see *The Report of the Committee to Examine the Patent System and Patent Law* (Cmnd. 4407) London, HMSO (1970) chap. 16 (referred to below as Cmnd. 4407)
22. S. G. Sturmey *The Economic Development of Radio* London, Duckworth (1958) p. 275
23. Cmnd. 4407 (see n. 21) p. 147
24. G. Becker *Human Capital A Theoretical and Empirical Analysis with special reference to education* New York, Columbia UP (1964) chapter 2
25. J. E. Tilton op. cit. p. 81
26. J. L. Enos op. cit. pp. 228–9
27. K. Arrow and W. M. Capron op. cit.
28. *The Brain Drain: Report of the Working Group on Migration* (Cmnd. 3417) London, HMSO (1967)
29. See H. G. Grubel and A. D. Scott 'The International Flow of Human Capital' *American Economic Review* (1966); B. Thomas 'Brain Drain Again' *Minerva* (1967); and correspondence between H. G. Johnson, A. D. Scott and B. Thomas in *Minerva* (1967) and (1968). All reprinted in M. Blaug (ed). *Economics of Education* vol. 2, Harmondsworth, Penguin (1969)
30. H. G. Grubel and A. D. Scott op. cit.

PART 2

Channels for Technological Change

6 The Government's Role

In this part of the book the main channels for technological change are considered. This chapter deals with the role of government-financed R & D, whether undertaken intra- or extramurally. The first section briefly outlines the importance and organisation of government-financed R & D, with special reference to the UK. The second section examines the economic basis of government involvement in R & D. The third section discusses the uses of social cost–benefit analysis in assisting government decision-taking. The last section describes the work of the publicly-owned National Research Development Corporation.

6.1 The importance of government-financed R & D

In most advanced countries government support of R & D is extensive. Table 6.1 provides a breakdown of R & D activity in some OECD countries by source of funds and sector of performance. There are of course considerable problems in making international comparisons of this nature and the data must be treated with care. Nevertheless, the figures do give some indication of broad magnitudes and indicate the fundamental importance of government finance : in six of the ten countries 50 per cent or more of the total R & D effort was financed from this source.

There are however variations in the extent to which government undertakes R & D *intramurally*. For example, over a quarter of all R & D in France was carried out by the government, whereas in the US the figure was less than a sixth. The determinants of the relative size of the government's contribution to total R & D budgets are many and varied and will, *inter alia*, depend on the force with which some of the arguments outlined in section 6.2 can be applied.

Table 6.1
(% of total)
Total R & D Expenditure in some OECD countries

Country	Year	Source of funds (%)[a]			Sector of performance (%)[a]		
		Government	Industry[b]	Other[c]	Government	Industry	Higher Education
Canada[d]	1969	61.8	30.2	8.0	21.0	50.2	28.8
France[e]	1969	63.6	30.0	6.4	26.2	58.0	15.8
Germany[f]	1969	42.0	57.6	0.4	15.9	65.5	18.6
Italy[f]	1969	51.7	46.9	1.4	23.6	51.3	25.1
Japan	1969	28.1	60.1	11.8	11.2	61.7	28.1
Netherlands[f]	1969	40.7	57.7	1.6	21.2	61.2	17.7
Norway[f]	1969	61.1	34.9	4.0	21.8	44.1	34.1
Sweden[g]	1968	29.0	69.8	1.2	na	na.	na
United Kingdom[d]	1968	50.6	45.2	4.2	24.5	67.2	8.3
United States[e]	1969	57.3	38.0	4.7	16.8	70.3	12.9

[a] The classifications adopted above are somewhat although not substantially different from those used in the source text
[b] Includes nationalised industries
[c] Includes higher education, foreign sources, etc
[d] Does not include data for social sciences
[e] Includes data for law
[f] Includes data for law, humanities, education and arts
[g] Does not include data for social science

Source: *UNESCO Statistical Year Book, 1971* New York, UNESCO (1972) tables 3.7 and 3.8

These arguments are likely to be particularly relevant in the fields of atomic, space and defence R & D, which account for a major part of all R & D carried out in some countries. The US and UK spent about 62 and 40 per cent respectively of their total public and private R & D budgets in these fields in 1965.[1]

Table 6.2 provides a breakdown of government expenditure between intra- and extramural work in the UK. It can be seen that the amount of intramural work varies considerably between categories of R & D. For example, most government-financed aerospace R & D is undertaken in industry, while nearly all atomic energy R & D is done by the government itself. To some extent this

Table 6.2
Gross government expenditure on scientific R & D in the UK, 1970–71
%

	intramural	extramural		
		Universities and further education	Private industry and public corporations	Other	Total extramural
Defence	40.0	0.5	53.3	6.2	60.0
Civil					
Research Councils and other science grants	54.6	16.5	1.3	27.5	45.4
Aerospace	14.2	0.2	82.6	2.9	85.8
Atomic energy	94.4	0.6	3.2	1.7	5.5
Industrial services	51.8	2.6	45.0	0.7	48.2
Agriculture Fisheries Forestry	70.8	7.2	1.9	20.0	29.2
UK universities	—	100.0			100.0
Other	78.0	1.7	12.7	7.6	22.0
Total	43.1	30.0	26.7	13.9	56.9

Source: *Research and Development Expenditure* (Central Statistical Office) London, HMSO (1973) table 6E

division of labour may be determined by the relative efficiencies of the government and the private sector in carrying out certain types of R & D work. These may differ for several reasons. Firstly, in some R & D there may be economies to be derived from concentrating resources in a unit the size of which is bigger than that which most or all private firms could sustain individually. This argument is likely to be particularly forceful in agriculture.

118   CHANNELS FOR TECHNOLOGICAL CHANGE

To finance such a programme in industry might lead to the uneconomic splitting up of the work between firms. (This disadvantage would of course have to be weighed against any additional disadvantages that *might* arise from carrying out the work in government rather than in private institutions – see section 6.2). Secondly, an intramural programme may provide greater continuity for some R & D that would otherwise be subject to the fortunes of individual private companies. Thirdly, government may already have particular types of expertise that industry does not have (see below). Secrecy considerations may also play some part, although as shown in the table the majority of defence work, in which issues of security are very important, is carried out extramurally.

It must of course be remembered that the allocation of government R & D finance is a political as well as an economic process. Pressure groups and individuals within ministries, firms and government R & D institutions will all be pressing their own claims, which are unlikely to be related to economic issues alone. Current allocation of resources between intra- and extramural work may be the result of historical forces. For example, the Atomic Energy Authority (see below) is currently running down its main nuclear programme for which it was originally set up and is thereby releasing manpower for other work. Much of this manpower however is not being released on to the open market but is being retained within the AEA. The authority is now undertaking contract work for industry, some of which is unrelated to nuclear R & D.

Table 7.1 on p. 157 provides a breakdown of government spending in manufacturing industry, which constitutes most of its extramural work. Two features of this breakdown are particularly prominent. Firstly, only a few industries derive a substantial proportion of their R & D funds from government. The aerospace figure of course stands out here and is a reflection of both the civil and the military involvement of the government. Secondly, government R & D expenditure in industry is concentrated in only two industries, aerospace and electrical engineering, and particularly the former.

6.1.1   *The organisational framework of government R & D*

The division of responsibilities between UK government departments in 1971–72 for R & D as measured by expenditure is given in table 6.3. (The 1974 Spring election which led to a change of government also led to some reorganisation in government departments; however the latest available statistics refer to the old departmental structure.)

Table 6.3
Government R & D expenditure (estimates) by department 1971-72

	Amount (£m.)	% of total
Ministry of Agriculture, Fisheries & Food	6.2	1.0
Department of Agriculture & Fisheries for Scotland	6.5	1.0
Ministry of Defence	259.3	40.2
Department of Education and Science	109.5	17.0
Department of Environment*	33.2	5.1
Overseas Development Administration	3.5	0.5
Health Departments	10.9	1.7
Home Office	2.5	0.4
Department of Trade and Industry†	205.0	31.8
Miscellaneous	8.9	1.4
Total	645.5	100.0

* Includes expenditure on R & D laboratories and accommodation for other departments

† The R & D functions of this department are now largely the responsibility of the Department of Industry.

Source: First Report of Select Committee on Science and Technology, (HC 237, Session 1971/72) London, HMSO (1972) p. xxxii

Three departments, the Ministry of Defence and the Departments of Industry* and of Education and Science dominate the picture. The functions of the first two overlap in a number of fields. The overlap is particularly strong in the aviation field. (The old Ministry of Aviation was absorbed into the Ministry of Technology – one of the Department of Industry's predecessors – in 1967.) As a result there are a number of sharing arrangements between the ministries including a degree of common financial supervision.

It can be seen from table 6.2 that about two-fifths of the defence R & D, which includes that spent by the Department of Industry, was done intramurally. This work is undertaken primarily in the defence research establishments of which there are twenty-six. Eight of these, concerned principally with aircraft come under the Department of Industry. The *civil* R & D activities of the Depart-

* This department was formed after the splitting up of the functions of the former Department of Trade and Industry, following the 1974 Spring election.

ment are heavily concentrated on three projects – Concorde, the RB 211 engine and reactor R & D. These three account for over sixty five per cent of the Department's total civil programme.[2] The main bulk of its intramural activity is undertaken by the Atomic Energy Authority (AEA) and the industrial research establishments (IREs).

The AEA was established in 1954 to 'produce use and dispose of atomic energy and radioactive substances and to carry out or commission research into matters necessary to support these activities'.[3] Since 1965 it has also had the power, with government approval, to undertake non-nuclear work, some of which is on contract from industry or government departments. This currently amounts to just under ten per cent of its civil R & D programme of £67m. The Authority hopes to increase this work as the atomic energy programme continues to run down.

The industrial research establishments divide into two groups. Firstly, there are the multi-purpose IREs – the National Physical Laboratory, the National Engineering Laboratory and the Warren Spring Laboratory. Their work covers research that has wide applicability. For example, two-thirds of the NPL's work is concerned with standards, and the Warren Spring Laboratory is involved in research into atmospheric pollution. The second category of IREs consists of the two much more specialist units: the Safety in Mines Research Establishment and the Laboratory of the Government Chemist. In total, the IREs account for about £13m of the Department's R & D budget, as shown in table 6.4.

Table 6.4
Gross expenditure, 1970–71 industrial research establishments*

	Expenditure (£m.)
National Physical Laboratory	6.0
National Engineering Laboratory	3.2
Warren Spring Laboratory	1.3
Safety in Mines Research Establishment	0.9
Laboratory of Government Chemist	1.0
Total	12.5

* These establishments received £0.8m. in receipts in 1970/71
Source: Third Report of Select Committee on Science and Technology (HC 302 Session 1971/72) London, HMSO (1972) p. x

Both the AEA and the IREs have been the subject of criticism on the grounds that they are not always sufficiently geared to the needs of industry. It is difficult to know however whether this criticism, if true, is inherent in the nature of these organisations or whether administrative innovations can overcome it.[4] Certainly there is no lack of suggestions of the latter nature.

At the more fundamental end of the R & D spectrum, the government finances five semi-autonomous research councils through the Department of Education and Science. Two of these, the Science and the Social Science Research Councils, have extensive responsibilities for financing postgraduate education. The other three are principally research bodies. Expenditure estimates for the five research councils are given in table 6.5.

Table 6.5
Expenditure (estimated), research councils, 1971–72

	Expenditure (£m.)
Agricultural Research Council	18.7
Medical Research Council	23.0
Natural Environment Research Council	15.8
Science Research Council	55.7
Social Science Research Council	4.1
Total	117.3

Source: *First Report of Select Committee on Science and Technology* (HC 237, Session 1971/72) London, HMSO (1972) p. xvi

Little work has been done on the effectiveness of the research council system although a review by the Council on Scientific Policy in 1971 concluded that it was 'essential for the best administration and conduct of this area of government sponsored activity'. It opposed the transfer of the research councils to particular departments although it did see a need for improving the links between the councils and government departments.[5] Each of the departments mentioned in table 6.3 receives its own vote. Because of the possibility of lack of co-ordination in the framing of the R & D policies of individual departments, the Select Committee on Science and Technology has recommended that a post of Minister for Research and Development be established.[6] This minister would have his own vote and would examine and approve all government R & D. This recommendation however has not so far been accepted by the government. The previous (Conservative) government argued that the R & D activities of individual departments could not be separated from their other objectives. It also argued that machinery for ensuring co-operation and co-ordination between departments already existed.[7]

6.1.2 *The 'customer–contractor' relationship*

In 1971 the Central Policy Review Staff, the government's 'think tank', led by its (then) Director, Lord Rothschild, recommended that government-financed applied R & D in the UK should be undertaken on the basis of a customer–contractor relationship: 'The customer says what he wants; the contractor does it (if he can); and the customer pays'.[8] These proposals came in for considerable criticism, from scientists and others.[9] The criticisms were related particularly to the transfer of decision-making from the research organisations themselves and the scientists actually undertaking the research to the departmental committees which are necessarily more remote from the laboratory.

Some contractor organisations were already acting on this basis in a limited way: for example, the AEA and some of the IREs were accepting work on a contract basis. Several 'customer' government departments on the other hand have now been reorganised on their R & D side to reflect this policy. For example the Department of (Trade and) Industry established a series of 'requirements boards' covering a number of fields on which customer interests in the Department, in industry and elsewhere and, where appropriate, the contractors are represented. The change that this new organisation is intended to bring about, for example in relation to the formulation of the IREs' research programme, has been described by the Chief Scientist at the Department:

> Each IRE will end up by having a portfolio of requirements which will have arrived through different requirements boards. Rather than the present situation, where the programme is more or less invented within the IREs and then endorsed by a process of advisory committees and ultimately by myself, the programmes will be arrived at by bringing together different requirements which have been examined by competent Boards. In some ways the IREs will be competitors to other people who carry out

research and even, once in a while, competitors one to the other.[10]

Perhaps the most contentious implications of the new system were those for three research councils (Agricultural, Medical, and Natural Environment). Part of the budgets of these research councils has been transferred to the appropriate customer departments to finance part of the latter's commissioned research. Although no conditions are to be placed on the use of the money so transferred, it is expected that it will be used to commission applied research work from the Councils. (The Science and Social Science Research Councils are for the present left out of these new arrangements.) A surcharge may be added on to the contract price to allow contractors an element of free-ranging work.

The widespread adoption of the customer–contractor principle in government-financed applied research is likely to lead to greater clarity in the roles played by different individuals and institutions in R & D activity and to more specific formulation and assessment of requirements, even though the same personnel as under the earlier system may still be involved. In this respect, the government is relying on *contractual* arrangements to bring out a better allocation of R & D resources. However, we do not yet know what the effects of introducing an arrangement of this kind will be on creative research work. Much more study needs to be devoted to analysing the effects, if any, of different administrative and financial arrangements on creativity. Furthermore, we are not yet in a position to judge how accurately the Requirements Boards will reflect the true needs of the 'customers', ie society at large.

6.2 The economic basis for government-financed R & D

The economic case for government-financed R & D in capitalist economies is usually based on an analysis of the inadequacies or imperfections of the market mechanism in securing a socially optimal allocation of resources to R & D and associated activities. A possible definition of a situation where such an optimum may be said to exist is one in which it is impossible to make any individual better off without making someone else worse off. Such a situation – known as Pareto optimal in welfare economics – does however imply several important value judgements about the role of the individual in determining his consumption and production patterns.[11] Given these values and some fairly restrictive assumptions, it can be easily demonstrated* that a perfectly competitive system will provide such an optimum. (This may not necessarily imply a system of private enterprise; it may still be possible to implement the same decision rules derived from Paretian analysis in a socialist economy.) Two of the restrictive assumptions used in this analysis are particularly inappropriate in considering the allocation of resources to R & D : firstly, that there are no externalities, ie there is no divergence between private and social costs and benefits; and secondly, that perfect information exists, ie there is no uncertainty. (It may be argued that this is an assumption that is an integral part of perfect competition.) These two issues are dealt with in turn below.

6.2.1 The nature of R & D

The very nature of R & D is likely to give rise to a divergence between private and social costs and benefits. As we have seen (p. 24), one of the characteristics of the end product of R & D – information – is that it is not wholly appropriable by the institutions that generated it. This is likely, in turn, to mean that others will also receive benefits without paying costs. As was shown in chapter 1, basic research results are likely to be particularly difficult to appropriate : research workers will, after a time lag, be able to utilise the results of others in their own work by reading learned journals and through professional contacts. They will not have to buy them in the market, although they may incur costs associated with the transmission of the knowledge, eg the purchase of books, etc. Other information apart from basic research findings may also be inappropriable. For example, the results of development work will usually be expressed in a company's product. An outsider may be able to obtain useful information merely by buying and then analysing the goods in question. His ability to do so will depend on the precise way in which technical progress is embodied in the

---

* Most intermediate micro texts provide an analysis of the conditions under which Pareto optimality is achieved. See, for example, W. Baumol, *Economic Theory and Operations Analysis* (3rd edn), Englewood Cliffs, Prentice Hall (1972) chap. 16.

goods in question. In these contexts, the incentive for private enterprise to engage in basic research or to produce other information that can be appropriated by others without charge will be greatly reduced, and society may not therefore devote sufficient resources to these activities. Private marginal costs may still be equated with private marginal revenues. But the private, profit-maximising 'output' will be less than the social optimum.

It seems clear therefore that private enterprise is likely to under-invest in R & D because of the nature of the commodity produced by this activity. Some attempt is made of course through patent laws and industrial secrecy to overcome this to some degree (chapter 2), and no doubt other legal measures could be taken to establish property rights, but such moves are unlikely to provide more than a partial remedy. In any case, because information is a public good – consumption by one person does not diminish the supply available to others – it is doubtful whether appropriation and the consequent ability to charge users would be an optimal policy, since the social cost of using *existing* information is virtually zero (although there are costs of applying it). It may also be argued that free cross-fertilisation of ideas between scientists unhindered by economic barriers is an essential prerequisite for the efficient pursuit of research. For many types of R & D therefore, social returns are likely to exceed the private returns. The anticipated private returns may still be large enough to induce an enterprise to undertake some research (but this may mean that it is at a competitive disadvantage since it has incurred costs not incurred by its rivals).

Where research results in a saleable commodity a firm may be better able to recoup its R & D costs – Weisbrod uses the illustration of polio vaccine.[12] While the research knowledge that led to this development is freely available, it can be *applied* only through a vaccine for which payment must be made. Thus where the investment in research and in the application of that research are considered jointly, an optimal level of research activity may result. However as Weisbrod has pointed out, it may not be clear to a firm embarking on a research project that the latter will necessarily result in a saleable commodity. Furthermore, the problem of competing producers using the knowledge to produce their own vaccine still remains. Even if the expected private returns are sufficient to induce a firm to undertake a given programme of work, this may not be the socially optimal solution, since the firm concerned may, through patent rights etc, restrict the utilisation of the R & D results so that other social benefits are not realised.

Social returns of an R & D programme may be far greater than the private returns accruing to the firm financing it, not so much because other firms can appropriate the knowledge without charge, but because the good or service in which that knowledge is incorporated is not traded in the market. Reduction in atmospheric pollution for example may be a social benefit for which a private manufacturer undertaking research in this field may receive little return in his own accounts. This stems essentially from the absence of property rights in pollution-free atmosphere.

Defence R & D is in a special category of its own as it raises special problems of security etc, although many of the basic principles are the same. Individuals benefit (or suffer) from such R & D independently of their own contribution; by the nature of the goods, no one can be excluded from its effects. Payment by 'consumers' is unlikely therefore to provide the socially optimal amount of defence R & D, with individuals attempting to obtain as much of a 'free ride' as possible, ie taking the benefits without paying the costs.

Scientific investigation might also be defended in terms of its own intrinsic merits, ie that it provides non-quantifiable benefits of improving the quality of life by increasing man's knowledge about the natural world.

At the begining of the chapter reference was made to the *discounted* value of social benefits. Individuals and society as a whole may however have different time preferences, so that even where in each individual time period the social and private benefits are equal, their discounted value is different. If the private discount rate is different from the social rate, as some economists have argued, then public support of R & D may be justified.[13]

The above discussion on the extent to which social costs and benefits may be inadequately reflected in private costs and benefits in a market system – and the effects of this discrepancy – should be qualified in two respects. Firstly, basic research may be an attraction to first class staff. Some firms, particularly the larger ones, may therefore undertake a limited amount of 'free ranging' research to provide this attraction, even though they may receive little return from the research itself. Secondly, while some research results may not be appropriated by individual firms, the latter may nevertheless obtain a competitive advantage by being first in the field, for example by shortening the gap between basic research on the one hand and development and exploitation on the other. There is no *a priori* reason however why either of these two factors should guarantee that the optimal level of spending will be achieved.

6.2.2 R & D and uncertainty*

A second reason why the competitive system may not provide an optimal allocation to R & D is the existence of uncertainty. As we have already seen (p. 22), the results of R & D are often uncertain, more so at the basic than at the development end of the spectrum. Consequently, under-investment in these activities is likely. Even if insurance against lack of success was possible, this would tend to create its own problems since it might well in turn reduce the firm's incentive to succeed.

The degree of risk involved in undertaking a particular programme will in part be determined by the structure of the firm concerned. The wider its technological base, for example, the more likely it is that it will be able to incorporate findings that were not originally anticipated. The width of a firm's technological base in turn is likely to be related to its degree of diversification. A large company is also able to reduce the degree of risk involved in an R & D programme by undertaking a range of projects each of which is comparatively small in relation to the programme's total size. One answer therefore to the problem of risk aversion by companies might be to alter the structure of industry. However, bigger combined units may create diseconomies in, say, management and production. They may also reduce the incentive to undertake research because of the increased concentration of firms.

There is also another slightly different problem associated with risk-bearing which may be very important in the context of R & D activity in a private enterprise economy. Some firms, because of their limited management capability or imperfect information, may vastly over-estimate the risks from R & D and under-estimate the benefits to be obtained from it. For example in investigations carried out in Britain in the late 1940s by the Board of Trade, a widespread

---

* We are using the term uncertainty here to include situations where probabilities can be assigned to particular outcomes.

ignorance of the possible value of scientific investigations was revealed.[14] External finance may be justified as a 'pump priming' exercise in cases of this kind, although other measures – eg of an 'educative' nature – might be an alternative possibility. (It is also interesting to consider the possibility that industrial R & D activity may act as a 'pump primer' on government-financed R & D. Investigations in industry may, for example, demonstrate the potential in social terms of a new product or process on which further R & D financed by government would be justified.)

An aversion to risk may also reduce the availability of venture capital necessary to undertake development projects, although this does not seem to be a very important consideration in Britain, at least for smaller projects (see p. 285).

6.2.3 *Implications for government*

The above arguments have provided some possible reasons why a competitive system of private enterprise may not provide an optimal allocation of resources to R & D and why other financing institutions may be necessary. Before looking at the implications of this for government R & D it is perhaps as well to draw attention to a danger that is often associated with this kind of discussion. It is very easy to slip into the error of arguing that because, for example, government is currently financing and/or undertaking an activity, its action is therefore justified in terms of the arguments outlined earlier. This may not follow. Nevertheless, there are few economists who would deny that even a perfectly competitive system is unlikely to provide a socially optimal allocation. Arrow, for example, has argued that 'for optimal allocation to invention, it would be necessary for the government or some other agency not governed by profit-and-loss criteria to finance research and invention'.[15]

Two points must be made on Arrow's conclusion. Firstly, there is a *whole range* of options open to a government to ensure that more R & D is undertaken than would otherwise be the case. It may *finance* work in other non-government institutions; it may undertake the work intramurally; or it may impose legal requirements on manufacturers which make it necessary for them to undertake certain kinds of R & D. This approach has been adopted in the control of pollution from exhaust fumes. It must also be remembered that in many areas the government acts as a major buyer of machinery and equipment. This in itself may ensure that some private R & D is undertaken. For example, Vosper Thornycroft's military hovercraft is a privately financed venture (see p. 240). The government will of course eventually pay for the R & D if the craft is successful, but the initial decision was a private, rather than a governmental one. However the uncertainty attached to securing government orders sometime in the future means that under-investment will probably still take place. Furthermore, it may be very difficult for a private firm to know precisely what is required by the government.[16] A government department may attempt to overcome these problems by placing *firm* orders for the finished product rather than by financing the R & D directly. Yet another alternative strategy to increase R & D activity could be implemented through the tax system. Governments must choose from these options. Unfortunately, little is known about their relative efficiencies.

Secondly and more fundamentally, Arrow's conclusion cannot be supported on an analysis of the shortcomings of the private enterprise system alone; as Demsetz has pointed out:

> Whether the free enterprise solution can be improved upon by the substitution of the government or other nonprofit institutions in the financing of research cannot be ascertained solely by examining the free enterprise solution. The political or nonprofit forces that are substituted for free enterprise must be analysed and the outcome of the workings of these forces must be compared to the market solution before any such conclusions can be drawn. Otherwise, words such as 'government' and 'nonprofit' are without analytical content and their use results in confusion.[17]

Demsetz's point carries considerable weight: shortcomings in one system may be remedied by introducing modifications but these modifications may themselves cause other problems which the original system did not have. Only when the total effects of providing adjustment have been analysed can an alternative policy be advocated. In fairness, however, it should be noted that Arrow does discuss some of the difficulties associated with government financing of R & D.

Political considerations, particularly those associated with the prestige of individuals or departments and with the relative strength of pressure groups, may play a major part in R & D decision-taking, and result in 'distorted' decisions.[18] Furthermore, assuming risks not carried by private enterprise also carries dangers for government departments and perhaps for individual civil servants. In some cases, Demsetz argues, 'the government is likely to behave towards risk in a much more risk aversive fashion than industry'. C. J. Hitch also made the same point in relation to American military programmes:

> In the government the riskiness of a research project is more of a deterrent than in the economy: the government official has to be cautious – the penalty for a striking failure is almost always greater than the reward for a striking success. So is the fact that the pay-off from research may occur only in the distant future: military research administrators normally rotate at three-year intervals and the political officials and Congressmen above them rotate fairly frequently too. In fact everyone above the level of the laboratory seems to have some good reason for demanding quick results . . . a circumstance unfavourable to basic inquiry.[19]

Government's *capacity* to bear risk may be greater, but this is no guarantee that it will have the commensurate skill in *identifying* those projects that should be supported. The record of government support for aircraft R & D is not encouraging in this respect.[20] Government intramural research may also suffer in efficiency terms because the discipline of the market is absent. Government laboratories may tend, for example, to utilise more costly equipment in the pursuit of their research, or to extend existing research programmes beyond what could be justified on economic grounds. On the other hand, the absence of direct commercial pressures may be conducive to creativity. Even if this were not the case it would be false to assume that government scientists do not work under pressures, albeit of a different kind from those present in private industry, that induce efficiency. Unfortunately again, very little is known about the relative efficiencies of government and industrial laboratories.

Government R & D contracts in industry may give rise to inefficiencies in the way in which the programmes are conducted. The cost plus contract for example – in which the government pays on the basis of costs incurred rather than on results – may lead to abuses by industry, even though it may be an effective device for shifting the risk from industry to government. Industry's incentive to control costs or to pursue the most rewarding avenues of work in a programme financed by government may be greatly reduced.

R. A. Solo for example found, in his study of the US synthetic rubber industry in the early postwar period, that a few companies with no research contracts from the government, no government money and relative freedom to exploit their own discoveries for profit were responsible for the major technological advances in that industry.[21] On the other hand, Jewkes's work points to many important advances achieved in government-financed work in independent research laboratories.[22] Much therefore may depend on the particular institutional framework in which the research is carried out. However, as Arrow has pointed out, the awarding of government contracts in industry often depends on past efficiency in executing contracts. Thus an incentive to efficiency remains, provided governments can correctly assess past performance, and provided the latter is closely correlated with future performance. The evidence on this second qualification however is not encouraging.[23]

Some attempt may be made to compromise between the efficiency and risk-bearing effects of government finance by providing shared contracts. For example, Hawker Siddeley and the UK government agreed to share the costs of developing the new short-haul airliner, the HS146 with the company undertaking to bear any costs of overruns.[24] Even here however (as this example shows) an initial plan to provide incentives may be overtaken by events, especially where costs and demand are difficult to estimate. (Hawker Siddeley now intends to stop work on the project, repaying the government aid it has received, because it sees no possibility of the project becoming profitable). Another example of this type of situation is found in the development of Concorde. Between 1968 and 1972 the British Aircraft Corporation (BAC) was responsible for about ten per cent of the total British investment in this project. BAC's (variable) profit on the project is based on a range of costs between 90 and 115 per cent of target costs and on the payload achieved between 18,000 and 25,000lb. Once costs exceed 115 per cent of the target, BAC can increase the amount of their profit only by increasing the payload. The Public Accounts Committee, which examined the whole Concorde programme, found that:

> In the event, actual costs have far exceeded 115 per cent of the target and it appeared to us that the arrangements were likely to have ceased to act as any significant incentive to keep costs down and, in fact, could encourage BAC to try to persuade the

Government to spend more money to improve performance so that the Company's profit would be greater.[25]

In a more general context the Committee also took the view that it was desirable that contractors should contribute towards the development costs 'as an encouragement to economy and to ensure realistic appraisals of the continuing commercial value of a project.[26]

In the case of Concorde, the problem has been one of controlling costs because BAC's profits are in any case going to be low. However in some cases of government-assisted projects company *profits* have been very high. This has often led to considerable government embarrassment and on occasions to the repayment of part of the profits.*

Another possible effect of government finance for R & D in industry is that it may act as a substitute for, rather than as a complement to, industry's own finance for R & D. Firms may use government funds to finance projects that they would otherwise have financed themselves from private sources. Two points should be made here however. Firstly, it would be virtually impossible to find out whether substitution is taking place by *asking* industrialists; secondly it may be possible by the terms of the contract to avoid some of this substitution. In the absence of circumstances referred to earlier – in which the use of the results from private R & D is restricted below the socially optimal level – substitution of government for private resources should be made only where the redistributive consequences are recognised and made explicit.

The only people who have undertaken any research in this field are Stigler and Blank, who have attempted to quantify how much substitution of this nature is taking place by comparing figures for a number of US industries for:

1) the ratio of QSEs in privately-financed R & D to total employment in firms undertaking no government-financed R & D; and
2) the ratio of QSEs in privately-financed R & D to total employment in firms undertaking government-financed R & D.

* One documented example of this kind of situation occurred in 1964, when Ferranti was forced to repay part of the profits it had made on the 'Bloodhound' missile contract. See the Second Report of the Public Accounts Committee, *Guided Weapons Contracts Placed by the Ministry of Aviation with Ferranti Ltd* (HC 183 Session 1963/4).

The outcome of their investigation,[27] which covered nearly all the 1564 manufacturing firms who reported their R & D activity in a national survey in 1953, is shown in table 6.6.

Table 6.6
Average ratio of engineers and scientists engaged in private research to total employment in US manufacturing firms, by industry, January 1952

Industry	Ratio of engineers and scientists engaged in private research to total employment in manufacturing firms*	
	Firms with government-supported research	Firms with no government supported research
	(%)	(%)
Chemicals	3.1	3.8
Petroleum	1.6	2.5
Electrical machinery	0.8	1.8
Motor vehicles	0.4	1.5
Aircraft	0.4	1.9
Other transportation equipment	0.2	0.8
Professional and scientific instruments (except photographic)	1.6	3.6
Photographic equipment and supplies	1.6	1.5
Food and kindred products	0.5	1.2
Textile mill products and apparel	0.4	0.9
Paper and allied products	0.7	0.7
Rubber products	1.1	1.4
Stone, clay and glass products	0.7	1.2
Primary metal products	0.6	1.2
Fabricated metal products	0.4	1.3
Machinery (except electrical)	1.0	1.7
All other manufacturing	0.5	1.1

* Excluding those firms whose ratios of engineers and scientists on public or private research to total employment exceeded 10 per cent.

Source: G. Stigler and D. Blank *The Demand and Supply of Scientific Personnel* New York, National Bureau of Economic Research (1957) p. 59

It is clear from the table that in nearly every industry the ratio of QSEs in private R & D to total employment is lower for firms undertaking government-financed R & D than for those who are not. At first sight this suggests that firms receiving government contracts are substituting public for private funds. As Stigler and Blank have pointed out however the results must be interpreted very carefully for several reasons. Firstly, the award of a government contract may lead a firm to reduce its own private R & D so that the management of both programmes can remain efficient. Secondly, there is a possibility that the sample of firms considered is strongly biased. For example, the second column in the table depends heavily on how firms that do no research at all are treated. This in turn depends on the size of firm considered. Thirdly, if larger firms have lower private R & D intensities anyway, and if most government contracts go to the larger firms, then the table need not necessarily show any substitution.

Government contract finance in industry may have other effects. For example, it may lead a firm to rely very heavily on this source, and to stop thinking in terms of 'commercial markets'. This shortcoming may have far-reaching effects, since even when military products are sold on commercial markets they usually need *some* degree of adaptation. On the other hand, government finance – either directly for R & D or via military orders – may enable initial high costs to be lowered. For example, Tilton has shown that usually a new semiconductor device is very expensive when it first comes on to the market. If the government buys for military purposes, this may enable firms to acquire production experience and thus lower their costs. The cost reduction may then permit a substantial market penetration. Without initial government orders this may not be possible.[28]

It is clear that the problems associated with government financing of R & D in a market economy are highly complex. They revolve around three basic decision areas. Firstly, the optimal allocation of society's resources to R & D and to different projects within the total R & D effort must be decided. Social cost–benefit analysis which is dealt with in the next section, may go some way to providing the basic material for these decisions, but the limitations of this technique are considerable. In any case the use of cost–benefit analysis in assessing all the alternatives would be a task of impossible magnitude. Secondly, some division must be made between the R & D that is likely to be privately sponsored and that for which it will be necessary or desirable to use government funds. Thirdly, decisions have to be reached on where government money is to be spent. How much should be spent intramurally and how much extramurally? How should the money be divided up *within* these categories? Which firms, if any, should be given contracts? What should be the terms of the contract? At present the attempt to answer these questions is very much on a hit-or-miss basis. However a review of past experience may enable certain faults to be avoided in the future.

6.3 Social cost–benefit analysis of R & D

If some idea could be obtained on an *ex ante* basis of the likely *social* rate of return on individual R & D projects, ie where *all* costs and benefits have been evaluated, then the allocation of public funds would be made easier. All projects which have a positive net present value (NPV) would be accepted. NPV may be defined as follows:

$$NPV = \sum_{t=0}^{T} \frac{B_t - C_t}{(1+r)^t}$$

where B is social benefit and C is social cost, both in year t. T is the life of the project and r is the given social rate of discount.* An alternative, although in the eyes of many economists less satisfactory, approach to the NPV method is to calculate the internal rate of return. This is done by setting NPV in the above equation equal to zero, and then solving for r. This rate of return can then be compared with the given social rate of discount. If it is higher, then the project will be commissioned. Where, however, a choice has to be made between projects because of, say, limited funds, the decision rules are rather more complex.†

Although the basic framework of cost–benefit analysis is relatively straightforward, there are a multitude of issues in connection with its theoretical basis and practical application that make an evaluation exercise much less simple than it at first appears. The valuation of costs and benefits, the treatment of risk and the choice of the social discount rate all raise difficult problems. Since the problems are of a general nature and do not relate specifically to R & D,

---

* The use of a discount rate does of course imply the value judgement that future benefits are less valuable than present ones. The constancy of the rate over time and its equal applicability to both costs and benefits are also assumed in the equation.
† The literature on cost benefit analysis and on the ranking of projects is very considerable; for a brief but thorough treatment see D. W. Pearce, *Cost Benefit Analysis* London, Macmillan (1971)

they are not dealt with in detail here although the examples given later illustrate some of the difficulties.

In cost–benefit analysis, both the costs and returns are estimated. Since costs are sometimes easier to estimate than benefits, a cost effectiveness approach is often used, in which only the costs of achieving a particular objective that may be expressed in non-monetary units, are estimated. Although this is a big step forward in comparison with the situation where no estimates of either costs or benefits are provided, its value is limited since it cannot provide an assessment of the objectives themselves.

The use of cost–benefit analysis in R & D has developed only very slowly, although interest is growing. For example, the Department of Industry and the Atomic Energy Authority jointly support a Programmes Analysis Unit. This unit has so far attempted about seventy evaluations.

Most of the published work to date has been done primarily on an *ex post* basis although some limited forward projections have also usually been involved. The best-known study to date is probably that by Griliches on the social returns to hybrid corn research in the US.[29] The basic method he adopted is relatively straightforward. Taking 1955 as the year in which the development was taken as closed, he estimated the total public and private expenditures on hybrid corn research. These costs were then cumulated and expressed as a capital sum as of 1955. Griliches then estimated the annual gross social returns from the research which he assumed to be approximately equal to the value of the consequent increase in corn output plus a price change adjustment (see below). He expressed past (ie up to 1955) and future returns as a perpetual annual flow. His third step was to estimate the additional costs of producing hybrid corn over the costs of producing the next best alternative. This was then subtracted from the annual gross social returns. He then estimated his returns as in table 6.7, using discount rates of 5 and 10 per cent.

These data imply an internal rate of return between 35 and 40 per cent (Griliches considers that even this is probably an underestimate). This is a very high return compared with most other investments. Griliches' valuation of social benefits was based on an estimate of the losses that would be incurred if hybrid corn 'disappeared'. This method of measurement is shown diagrammatically in figure 6.1. D is the demand curve. P_1 and P_2 are the old and

Table 6.7
Rate of return on hybrid-corn research expenditures as of 1955
($m.)

		$r = 0.05$	$r = 0.10$
(1)	Net cumulated past returns	4405	6542
(2)	Past returns expressed as an annual flow	220	654
(3)	Annual future gross returns	341	341
(4)	Annual additional cost of production and research	93	93
(5)	Total net annual returns: (2) + (3) − (4)	468	902
(6)	Cumulated past research expenditures	63	131
(7)	Rate of return: 100 × (5)/(6)	743	689

Source: Z. Griliches 'Research Costs and Social Returns: Hybrid Corn and Related Innovations' *Journal of Political Economy* (1958) (© University of Chicago Press)

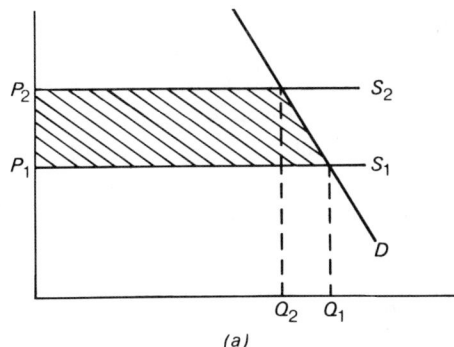

(a)

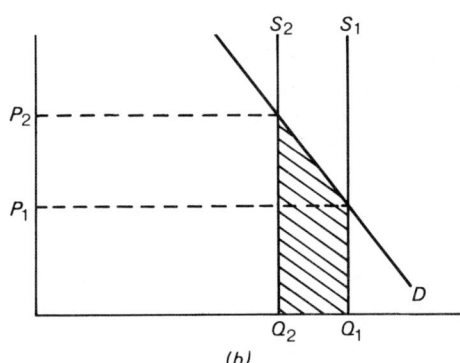

(b)

FIGURE 6.1(a) and 6.1(b)

new equilibrium prices (ie with and without hybrid corn) respectively. Q_1 and Q_2 are the old and new equilibrium quantities. He assumes, on the basis of a published estimate, that demand has a price elasticity of 0.5. However, because no similar estimates are available for supply elasticities, he looks at the effects of assuming infinite elasticity (figure 6.1(a)) and zero elasticity (figure 6.1(b)). The loss that would result if there was no hybrid corn would be the shaded areas. In figure 6.1(a) the loss would be made up of two factors: the increase in cost of producing the new quantity – Q_2 (P_2-P_1) – and the loss in consumer surplus caused by the price rise from P_1 to P_2: $\frac{1}{2}(P_2-P_1)(Q_1-Q_2)$. Where zero supply elasticity is assumed (figure 6.1(b)) the loss would be measured by the loss in production at the old price P_1: P_1 (Q_1-Q_2), and the loss in consumer surplus caused by the price change: $\frac{1}{2}(P_2-P_1)(Q_1-Q_2)$. In the context of Griliches's figures the second loss would be approximately 13 per cent bigger than the first.

Very little of the social returns were appropriated by the private producers, either of the seed or of the corn. This research was therefore a classic case of the social returns being very much greater than the private returns. Griliches however argues that this of itself is not sufficient to justify public investment:

> We must ask not only whether social returns are higher than private . . . but also whether the private rate of return is too low, relative to returns on alternative private investments to induce the *right* amount of investment at the *right* time [Griliches' italics].

At the same time however it is very unlikely that the relationship between the private rates of return on different projects would exactly correspond (even though they might be lower) with the structure of social rates of return on the same projects. Only if this was so could industry be relied on to provide a socially optimal allocation of resources.

A more recent cost–benefit study by Weisbrod analysed the rate of return from the development of vaccines for polio.[30] This study is particularly interesting as it serves to emphasize some of the difficulties likely to be met in an exercise of this nature. He estimated the internal rate of return, r by solving the following:

$$\sum_{t=0}^{T} \frac{R_t - B_t(N_t - W_t) - V_t}{(1+r)^t} = 0$$

where R is research costs, B is the benefit per case of the disease prevented, N is the number of cases occurring in the absence of the vaccine, W is the number of cases occurring with the vaccine, V is the cost of applying the research fiindings, t is a particular year and T is the year beyond which the value of the variables 'is asserted to be irrelevant'. On the basis of different estimates and assumptions, Weisbrod arrived at a range of internal rates of return which varied between 4 and 14 per cent, the most likely rate being around 11–12 per cent.

The basic problems Weisbrod faced were three-fold. Firstly, he had to estimate what would have happened in the absence of the vaccine. This involved a number of assumptions. Alternative assumptions were tried; thus the problem was at least made explicit. Secondly, he had to decide what to include and what to exclude on the costs side. This problem is discussed more fully below. Thirdly, and far more difficult to resolve, he had to value the benefits from polio vaccine. Weisbrod's basic approach here was to estimate the increased production via the present value of expected earnings resulting from the lower death and disability rates, and the reduced costs of treatment for people who would otherwise have died or been permanently impaired. How to value life is of course an extremely complex problem. The goods and services produced by the individual during his lifetime are only one aspect – some would say a very small aspect – of the social value of that individual's life. (How would the value of greater longevity for retired, crippled or mentally retarded people be assessed?) However, if the limitations of the study are clearly set out, a cost–benefit analysis on this basis may still be a useful aid in decision-taking.

Valuing forgone production is nevertheless still a problem. While the use of marginal productivity analysis to assume that a worker's wage is equal to the value of his marginal product, is a simple way of meeting the problem it begs a number of questions about the underlying theory. Furthermore many services, eg of housewives, are not marketed at all. There may be ways around this particular problem. The Programmes Analysis Unit for example attempts to value the housewife's time on the basis that it is at least equal to what she could have earned in alternative employment.[31] It must also be remembered that while death reduces output it also leads to reduced consumption – that would have occurred both during the working life and in retirement. The arguments here are in many ways related to those over the losses and gains from the brain drain (see pp. 108–10). The valuation of life is only one of the many areas in which cost–benefit practitioners face difficult issues. Valuing amenity, unpolluted air and uncongested streets are just a few other benefit areas that present similar problems.

One of the difficulties involved in both the Griliches and Weisbrod studies is deciding what to include in the cost calculations. Hybrid corn itself was a highly successful development, but what about the *un*successful projects undertaken in related fields? Should *they* be included in the hybrid corn research costs? How should we deal with unsuccessful projects in apparently unrelated fields? There are grounds for arguing that, in a broad sense, hybrid corn would not have been invented had it not been for the fact that R & D resources were spread over a wide range of fields. If this argument is accepted then, as Griliches points out, the only rate of return that is meaningful is one for R & D *as a whole*.

Similar problems faced Grossfield and Heath in their estimates of the returns to the NRDC's expenditure on a potato harvester.[32] They did in fact include the costs of developing an unsuccessful machine, which had a very similar purpose to the one that proved to be successful. It is interesting to see that one of their conclusions is crucially dependent on whether or not this cost is included. The NRDC's and National Institute of Agricultural Engineering's returns substantially exceeded the direct development expenditure of the two bodies if the unsuccessful project was excluded from the calculations, but the account becomes unprofitable if it is included. On the benefits side, a research project may give rise to highly successful spin-offs and a decision has to be made on whether or not to include them in an evaluation of the original project.

The *ex post* approach, while providing useful methodological and factual material which may *inter alia* be useful in providing a checklist for new projects, nevertheless avoids one of the crucial difficulties of *ex ante* assessments of R & D – that of the uncertainty of the outcome in both commercial and technical terms. This problem may still be very important even in development where the degree of certainty is usually thought to be at its greatest. Grossfield and Heath, for example, concluded in their study :

> We are clear that no analyst could reasonably have foreseen the unfolding of this innovation as it occurred, and that it would be a lucky shot indeed to have estimated at the appropriate times

in the past, the present value of the prospective stream of net social benefits.[33]

There are a number of interrelated problems associated with uncertainty in evaluating R & D. Firstly, the technical outcome may itself be uncertain. As Nicholson has pointed out, it may not be just a question 'of the reliability of estimates of cost or of time to completion . . .'. The uncertainty here is whether the objective can be attained 'regardless of input or research resources'.[34] Some system of estimating probabilities of technical success must be employed to overcome this difficulty. Estimates must also be made of the likelihood of a technical advance being overtaken by a further advance, since this will affect the discounted value of benefits. A technical advance may also be rendered redundant by changes in the political environment. Military weapons, for example, will be particularly prone to this uncertainty. Secondly, the commercial outcome may be uncertain. Technical success is no guarantee of commercial success: the unsuccessful harvester in the Grossfield and Heath study was awarded a silver medal by experts for its technical characteristics.

One major determinant of the commercial success of R & D is of course costs, and it is well known that these often escalate far beyond the original estimates, even in an area in which the technical outcome, in very broad terms, is predictable. Military programmes particularly seem prone to this kind of escalation.[35] On the civil side perhaps the best documented case of cost escalation in the UK is on the Concorde project.[36] Tables 6.8 and 6.9 give some idea of the increase in estimated costs and break down the sources of the increases. It can be seen that total costs have increased by a factor of six since the project's inception.

It can be seen that just over one-third of the total increase is due to changes in economic conditions. These changes are not closely related to the technical progress of the project and their effect on the commercial viability of the project is likely to be smaller than that of the other items. Even excluding these items, however, estimates have risen by a factor of five. The second major item causing the increase was 'Additional development tasks', which resulted from changes necessary for the aircraft to meet its specification. For example, in 1968 it was becoming clear that Concorde would not be able to carry the required 20,000lb payload from Paris to New York. As a result major redesign of the wings, fuselage

Table 6.8
Concorde: changes in estimated cost
(£m.)

	Total estimate	UK share	UK expenditure
November 1962	150–170	75–85	
July 1964	275	140	
June 1966	450*	250	45
May 1969	730	340	170 (to 31/3/69)
May 1972	1970	480	330 (to 31/3/72)
June 1973	1065	525	380 (to 30/4/73)

* A contingency of £50m. was also estimated at this time. The cost of work beyond the certificate of airworthiness was included in this and later estimates

Source: see table 6.9

Table 6.9
Concorde: breakdown of changes in estimated cost
(£m.)

Changes in estimates from table 6.8	Cause of escalation: Changes in economic conditions*	Programme slippage	Revision of estimates	Additional development tasks	Other adjustments
Nov. 1962 to July 1964 (£170–275m.)	18	—	47	40	—
July 1964 to June 1966 (£275–450m.)	34	—	38	103	—
June 1966 to May 1969 (£450–730m.)	107	—	58	115	—
May 1969 to May 1972 (£730–970m.)	83	26	22	70	39
May 1972 to June 1973 (£970–1065m.)	65	20	10	—	—
	307	46	175	328	39
			£895m.		

* Covers changes in the levels of wages and prices of materials and in exchange rates

Source: Sixth and Seventh Reports from the Committee of Public Accounts: *Development and Production of the Concorde Aircraft* (HC 335 and 353, Session 1972/73) London, HMSO (1973) p. vii

and engine had to be carried out. Situations arose in which, if the specifications were to be met, new modifications were essential. Consequently the choice was one of either abandoning the project altogether or allowing the increased estimate. The 'Revision of estimates' figure reflects the optimism of previous estimates. It is of course closely linked with the previous items. 'Programme slippage' costs result from the increases in time required to complete the development programme. These may in turn come about as a result, for example, of new design studies being necessary.

While the Public Accounts Committee was critical of certain aspects of the contractual arrangements between the government and BAC, it was not able to say whether estimates could have been made more accurately or whether the estimates could have been adhered to more closely. As far as government officials were concerned the Committee did not think that they could have made major savings on this project given the political constraints within which they worked. It is of course possible that industry and in some cases government employees may deliberately underestimate the costs of a programme so that the chances of selling it are improved.[37] It is however extremely difficult to provide firm evidence on this score.

Uncertainty in both costs and benefits means that a *range* of possible outcomes dependent on different assumptions will usually be necesary. This in turn will indicate which factors are most likely to affect the final outcome. (Even this approach however is unlikely to be of value in research that is of a very basic nature.) It is interesting that *no* range of possible outcomes was provided for Concorde, a project of an 'exceptionally speculative kind'. One of the problems with providing several cost estimates is that it makes the monitoring of the efficiency with which a programme is carried out more difficult.

A cost–benefit framework may be used to set a ceiling on R & D expenditure. The benefits derived from attaining specific results can be estimated, and the maximum amount of R & D that would result in a reasonable rate of return calculated. Whether the target could be achieved within this ceiling would be left to the judgement of those managing R & D.

The above discussion has illustrated some of the very considerable difficulties associated with cost–benefit analysis of R & D. As long as its limitations and assumptions are recognised and made explicit, it may provide a useful aid to the allocation of resources in some areas. However, it is unlikely to provide any guidance in the field of military R & D, a major part of total government effort, because of the impossibility of valuing 'benefits'. A more limited cost effectiveness approach however may still be of value here in aiding project selection.

6.4 The National Research Development Corporation

One body that has remained in existence throughout the postwar period without any change in its basic functions is the National Research Development Corporation (NRDC). This body is dealt with separately here because its activities reflect many of the issues and problems associated with the government financing of R & D generally. It has however been subjected to very little scrutiny by economists.

The NRDC was set up in 1948* to secure the development or exploitation of inventions derived from publicly sponsored research and also of inventions from private sources that are not being developed or exploited sufficiently or at all. To enable them to do this the NRDC was given the power to acquire and manage 'exploitation' portfolios, the main part of these portfolios being patent rights. The NRDC also has the power to promote and assist research aimed at satisfying specific practical requirements where that research is likely to lead to an invention or where it has already resulted in any discovery and where inventions of practical importance may result from continuing the research. All of these functions with the exception of the management of inventions from public sources are to be undertaken 'where the public interest so requires', but no guidance is given as to what the 'public interest' is.

The statutory objective of helping inventions from private sources was primarily aimed at helping the frustrated private inventor who was unable to proceed because of lack of public support. However as early as 1953–54 it was becoming clear to the NRDC that there was not a vast untapped supply of individual inventions with

* Under the 1948 Development of Inventions Act. The NRDC now operates under the consolidated 1967 Development of Inventions Act which repealed the 1948 Act, and the subsequent amending Acts of 1954, 1958 and 1965.

commercial potential waiting to be applied. In its Report for that year it stated:

> Our five years' experience therefore requires us to repeat and extend what we have said on this topic before. Outside the field of light engineering and instrument manufacture, the isolated individual rarely appears to have any serious contribution to make to the advancement of technology. In the various fields of physics, chemistry, biology, medicine and the non-mechanical branches of the engineering sciences, no meritorious proposals have reached us from such persons.[38]

It would be useful to know whether any inventions rejected by NRDC were subsequently developed by other means. The author knows of no published case study of a successful postwar innovation for which the NRDC refused funds, although 'refusal' is a difficult word to define precisely, since it is in part dependent on the inventor's position in the bargaining process.

However, although the NRDC's role in relation to private inventors has not developed as anticipated, it has become increasingly involved in the support of inventions *from industry,* even though it was not originally contemplated that industry would be a major source of undeveloped or unexploited new inventions or processes. Furthermore, its financial support is given primarily to larger firms and not to small units as might be expected. Table 6.10, which relates to projects in existence in the year 1968–69, provides a breakdown of NRDC assistance by firm size.

Table 6.10
NRDC support in industry: breakdown by firm size

Firm size (employees)	Number of projects (% of total)	Authorised expenditure on projects (% of total)
Below 25	24	2
25–100	32	7
100–500	20	20
Above 500	24	71

Source: NRDC *Annual Report and Accounts for 1968/69* (HC 327 Session 1968/69) London, HMSO (1969)

It is clear that NRDC's support for industry in value terms is largely geared to the bigger firm, even though in terms of the number of projects its main support is with firms of under 500 employees. It must be realised however that support given to a large firm may be to assist in the development of an invention that originated from an individual. The development of the hovercraft illustrates this possibility (see part 3).

On the financial side the NRDC is given the objective of so exercising its functions as 'to secure insofar as can be done consistently with the fulfilment of their purposes that the return to them from their activities shall be sufficient to meet their outgoings on revenue account'. No time limit within which this obligation must be fulfilled is given. The NRDC either has found difficulty in attaining this objective, or has interpreted it very liberally: the 1968–69 *Annual Report and Accounts* reported the first surplus on revenue account after nineteeen years of operation. (The surpluses have continued since that year.) The Corporation's deficit has been financed by borrowings from the sponsoring government department made available under successive Acts.

The powers of the relevant ministers are wide. Although under the Industrial Expansion Act, 1968, a ceiling of £50m. was imposed on advances to the Corporation, the Minister has power to cancel liability to repay part of the advances under certain broad circumstances and to give relief on interest payable on any advance. This latter power has been exercised continuously since first given in 1965.

The financial obligations of the NRDC are inevitably reflected in the terms on which finance can be offered. In 1967 the Corporation complained that there was a lack of recognition by industry of the nature of the risk that it was prepared to support and the terms it must attach to that support: 'We do not, as do Government Departments, provide subsidies or place development contracts: we invest money in industrial situations in the expectation that proportionate profit will accrue to us if the project succeeds but that we shall lose part or all of our investment if the commercial hopes for the project are not realised'.[39] The Corporation's financial assistance may take a number of forms (see for example pp. 299–300) including taking an equity share in a company.

In nearly all cases, NRDC finance is combined with private money. The way in which the Corporation's money is repaid also varies; it may be on sales of products embodying the invention, or on sales of a specified sector of a firm's business, or on a firm's profits over a given period. It is impossible to say how severe the

NRDC's financial terms in relation to other possible sources of finance are since each case varies in the degree of risk, technological complexity, etc. However, it is probably true to say that in relation to private venture capital companies, NRDC takes a more long term view of its investments.

The NRDC is not a completely independent body. With certain exceptions, most sponsorship activities must be approved by the Secretary of State for Industry, who also has the power to issue directions of a general character as to the exercise of the Corporation's functions. He also appoints the Board members. The importance of the Department's role is shown clearly in the decision in 1973 to close down Tracked Hovercraft Ltd, an NRDC wholly owned subsidiary (see below).

A brief summary of NRDC's work over its life up to 1971–72 is

Table 6.11
NRDC: a review of its activities, 1949–72

Submissions	
Public bodies	12,394
Private individuals and firms	15,522
Total	27,916
Submissions accepted	5,670
Development	
Expenditure:	
Projects authorised but not yet financed	30
Projects currently financed	173
Projects financing completed	476
Total	679
Income:	
Projects not yet income earning	234
Projects currently income earning	110
Projects which earned substantial income	37
Discontinued unsuccessful projects	298
Total	679
Licensing	
UK licence agreements completed	1,413
Current foreign licence agreements	84
Total	1,497
Income-earning inventions	704

Source: NRDC *Annual Reports and Accounts for 1971/72* (HC 397, Session 1972/73) London, HMSO (1973) p. 14.

given in table 6.11. Forty-four per cent of all 'submissions', ie inventions communicated to the NRDC, came from public sources. (It is not entirely clear from the source of table 6.11 what the precise meaning of a 'submission' is. In particular we do not know whether it refers to single patented inventions only or whether in some cases it covers groups of inventions). The table also shows the extent to which NRDC sifts these submissions: only about one in five have been accepted for exploitation and of those accepted only a small proportion have been the subject of licence agreements with industry: one in three in the 1950s. Less than half the inventions licensed have become revenue-earning, and probably only a handful of those that have earned any revenue have earned an adequate rate of return on capital. It is clear that its sources of revenue are concentrated on only a few projects. And even within the category of 'substantial' revenue-earners, the majority of income is derived from only a handful of projects. The largest revenue-earners are listed by the NRDC as follows[40] (the source of the invention is given in brackets). It is interesting to note that only one of these came from the private sector:

> Cephalosporin (MRC–Oxford University)
> Trio-iothyronine (MRC)
> Pole amplitude modulated electric motors (Bristol University)
> Hamster cell line (MRC)
> Potato planter (private invention)
> Dental cement (Manchester University)
> Selective weedkillers (ARC)
> Carbon fibres (RAE)

The first is by far the most important. It also illustrates the lengthy time scale that may be necessary to bring an invention conceived in a university laboratory to the commercial stage: although the first patent was filed in 1951, the first marketable products did not emerge until 1964. And of the five companies given the opportunity to collaborate with NRDC, only Glaxo now remains in the field.[41] A much larger number of projects have been discontinued without any or with only a partial return. Failures have arisen for a number of reasons. Under Section 8 of the 1967 Development of Inventions Act, the NRDC can apply for relief in respect of liability to repay advances used in connection with development of inventions or incurred in connection with research projects. In its 1966–67 *Annual Report* the NRDC specified the sixty-six inventions or research projects on which it was seeking relief, and the grounds on which the application was made. Thirteen of these involved sums over £10,000 and a further seven, over £50,000. The reasons for failure varied widely, as the following extracts show[42] (the amount of the relief applied for is shown in brackets).

Deakins gears (4,752)
During the initial stages of development studies an American patent search disclosed extensive prior art.
Acetylene production (£160,072)
It became apparent that this project, as originally conceived, demanded advanced engineering techniques that had not been developed. A more limited aim later adopted achieved improvements over contemporary practice insufficient to encourage potential licensees to undertake further development.
Bryce Smith's organo metallic compounds (£4,301)
No commercially viable use for these materials emerged.
Continuous processing of leather (£12,201)
The technical problems proved too ambitious and the potential market shrank to an uneconomic level during the course of the development.

One of the most publicised discontinued projects not mentioned above was the one related to tracked hovercraft which was undertaken by NRDC's wholly owned subsidiary, Tracked Hovercraft Ltd (THL).[43] NRDC had taken the view in 1969, two years after THL's formation, that the funds required to take the project to the stage where it could be shown to be commercially viable was beyond its own resources. By mid-1972 the NRDC had decided to close down the project if no further sources of funds were forthcoming either from government or elsewhere. The Department of (Trade and) Industry had already taken the exceptional step, under pressure from the Treasury, of telling the Corporation that no further grants to THL would be approved. Early in 1973 the government decided to wind up THL and to continue certain parts of the programme elsewhere. The Select Committee on Science and Technology strongly criticised the government's decision on the grounds that THL was a major world centre for a new technology and that it failed to recognise the long-term technological and commercial advantages of the work the company was doing. One of its recommendations was that the structure and powers of NRDC should be reviewed on the grounds that the financial constraints under which the Corporation operates may not always be ideal:

'while the NRDC system is effective in supporting R & D which can be quickly exploited it seems unable to cope with successful R & D with a major long-term potential'.[44] The Committee's view may well be justified; certainly it was clear that, given the NRDC's statutory obligations, the project could not proceed with Corporation finance. However, the validity of the Committee's judgement is of course crucially dependent on its optimistic assessment of the project's long-term potential for which it provides little evidence.

NRDC's involvement raises the question of whether its original decision to become involved in tracked hovercraft was the best one on all the evidence available in 1967. This seems difficult to accept in view of the fact that six years later in evidence to the Committee the Chairman of NRDC stated :

> I do not believe for one minute, as a businessman, that major sums of money could have been earned and that income could have been obtained without having a system properly running in the United Kingdom. To achieve this you had to spend a lot of money and spend it quickly . . . [and later] . . . there were many very, very strong arguments against the total spent which was potentially very large (he spoke in terms of 'hundreds of millions').[45]

Although NRDC had a change of chairman in 1969, it is very unlikely that the Corporation's view of the commercial prospects of the project had changed *so* drastically. In this context it is hard to understand the original decision.

Taking the 1949–72 period as a whole, the NRDC is in considerable financial deficit. Table 6.12 gives some idea of the cumulative position although it must be noted that the figures take no account of the effects of inflation, nor is any discount factor applied to allow for differences in the timing of outgoings and income. The figures also exclude interest payments.

Of course, many of the NRDC's projects will continue to earn a return in the future and some projects currently at the pre-commercial stage will also become revenue earners. Consequently if a valid assessment of NRDC's role up to, say, 1972 is to be made, some estimate of future returns, given that all NRDC's current commitments were completed, would be necessary.

It is likely that the outcome of such analysis would still show the NRDC in substantial overall deficit. What would the implication of such a result be? If the selection of a few good revenue-

Table 6.12
NRDC: cumulative outgoings and income, 1949-72
(£m.)

Outgoings	
Payments to external organisations and to associated companies by NRDC including its subsidiaries	
Current joint development projects	18.8
Subsidiary companies	8.1
Associated companies	1.6
Other projects	2.8
Other outgoings	
Administration costs and patents	12.0
Payments to inventors	3.2
Projects completed and discontinued	4.2
Fixed assets	0.9
Total	51.6
Income	
Income from patent holding according to source of invention	
Defence departments	0.7
Civil departments	0.6
MRC	9.3
ARC	0.6
Universities	5.4
Computers	0.7
Private inventors	0.9
Other sources	0.1
Income arising directly from investment in projects	
Levies from joint venture projects	7.7
Project and loan recoveries, other capital receipts, miscellaneous income	4.2
Total	30.1

Source: NRDC *Annual Report and Accounts for 1971/72* (HC 397, Session 1971/72) London, HMSO (1972)

earners also inevitably implies the selection of projects that do *not* pay off, so that an overall deficit arises, then there is a clear implication that it is not economic even to try selecting projects with commercial potential. Two crucial qualifications must however be added to this conclusion. Firstly, there may be wider social benefits that do not appear in NRDC's accounts, which if taken into account would alter the financial balance. Grossfield has pointed out, for example, that the NRDC often accepts submissions even 'though their commercial prospects are very poor and are unlikely to cover the corporation's costs :

> ... important factors work against increasing the stringency in selection ... Significant and well appreciated national advantages arise if a wide range of inventions is offered to industry. A statutory obligation requires active exploitation by NRDC of inventions derived from Government Departments. It is important also that the inventive source should be made to feel that its inventions are sympathetically and actively considered.[46]

The NRDC's attitude to such inventions may act as an encouragement to the creation of the big revenue-earning inventions, some of which may *not* pass through NRDC's hands. The danger of this argument however is that if the social benefits are not explicitly assessed, it may provide justification for an open-ended commitment to the public support of inventions. Secondly, NRDC may be experiencing a learning process in project selection; ie it may be getting better at identifying the revenue-earners. This is quite likely as the Corporation builds on its own experience, and may explain why it has been able to return a surplus on revenue account since 1968–69. (This surplus is however calculated *after* interest relief from the Department has been deducted. The cumulative surplus on this basis over the five years since 1968–9 is £863 thousand. Ignoring the interest relief however changes this surplus into a deficit of £6.6m.) It must be remembered however that, while it may *now* be desirable for the NRDC to continue, on the grounds that its previous losses are 'bygones', it does not necessarily follow that it would be economically desirable for another country to set up an NRDC-type body from scratch if it cannot 'short-circuit' this learning process.

6.5 Conclusions

The government plays a very important role in the financing of R & D in most market economies. Governments as a whole and individual departments within those industries vary considerably however in the extent to which the work they finance is conducted intramurally. Most government-financed R & D, in UK and US industry at least, is concentrated in a few industries, particularly aircraft. In the UK there is a wide diversity of administrative arrangements for the financing and control of government-financed R & D.

It is unlikely that the market will provide a socially optimum allocation of resources to R & D, principally because of the intrinsic nature of the results of the latter, and of the risk involved. However, even if this is accepted, it does not necessarily follow that government should become involved, since such involvement may itself cause additional problems which more than outweigh those that it is intended to solve.

Social cost–benefit analysis does at first sight appear an attractive way of solving the intricate problems of allocating resources within R & D. However, there are massive difficulties in identifying, quantifying and then valuing the costs and benefits involved. In some areas the technique may just not be usable. *Ex ante* evaluations, which are necessary if cost–benefit analysis is to aid decision-taking, face the problem of forecasting, a problem that is particularly acute in the R & D field. Nevertheless, a cost–benefit approach may in some cases act as a useful aid to decision-making if its limitations are acknowledged.

The NRDC's experience in the UK provides a useful insight into the problems of sifting new ideas. While its overall deficit may raise the question of whether it should ever have been formed (even if its demise would mean that certain products with very great potential would not get exploited), it must be remembered that loss-making activities may be justified on wider social grounds or on the grounds that a learning process is taking place which will eventually produce benefits that more than offset past losses.

NOTES on Chapter 6

1. *The Overall Level and Structure of R & D Efforts in OECD Member Countries* Paris, OECD (1967) p. 58
2. Select Committee on Science and Technology *Fourth Report* (Session 1971/72, HC 308) London, HMSO (1972) p. x
3. Select Committee on Science and Technology *Minutes of Evidence* (HC 239 (ii) Session 1971/72) London, HMSO (1972) p. 24
4. Select Committee on Science and Technology *Second* and *Third Reports* (HC 294 and 302, Session 1971/72) London, HMSO (1972)
5. *The Future of the Research Council System* (Cmnd. 4814) London, HMSO (1971) p. 19
6. Select Committee on Science and Technology *First Report* (HC 237, Session 1971/72) London, HMSO (1972) p. iv
7. *Framework for Government Research and Development* (Cmnd. 5046) London, HMSO (1972)

8. *A Framework for Government Research and Development* (Cmnd. 4814) London, HMSO (1971) p. 3
9. See for example 'The Choice and Formulation of Research Problems' *Minerva* (1972)
10. Select Committee on Science and Technology *Third Report* (HC 302, Session 1971/72) London, HMSO (1972) p. xx
11. See for example S. K. Nath *A Reappraisal of Welfare Economics* London, Routledge & Kegan Paul (1969)
12. B. Weisbrod 'Costs and Benefits of Medical Research: A Case Study of Poliomyelitis' *Journal of Political Economy* (1971)
13. P. D. Henderson 'Notes on Public Investment Criteria for Public Enterprises' *Bulletin of Oxford University Institute of Statistics* (1965)
14. P. S. Johnson *Cooperative Research in Industry* London, Martin Robertson (1973) p. 36
15. K. Arrow 'Economic Welfare and the Allocation of Resources to Inventive Activity' in National Bureau of Economic Research *The Rate and Direction of Inventive Activity* Princeton: Princeton UP (1965)
16. M. Peck and F. M. Scherer *The Weapons Acquisition Process* Cambridge, Mass., Harvard UP (1962)
17. H. Demsetz 'Information and Efficiency: Another Viewpoint' *Journal of Law and Economics* (April 1969)
18. For a detailed account of the way in which political influences had an important effect on a research programme see Nevil Shute *Slide Rule* London, Pan Books (1968), especially pp. 100ff
19. C. J. Hitch *The Character of Research and Development in a Competitive Economy* Santa Monica, Rand Corporation (1958)
20. R. Caves et al. *Britain's Economic Prospects* London, Allen & Unwin (1967) pp. 474–6
21. R. A. Solo 'R & D in the Synthetic Rubber Industry' *Quarterly Journal of Economics* (1954)
22. J. Jewkes et al. *The Sources of Invention* London, Macmillan (1969)
23. M. Peck and F. M. Scherer op. cit. p. 374
24. *The Times* (30 August 1973)
25. Committee of Public Accounts, *Sixth* and *Seventh Reports* (HC 335 and 353, Session 1972/73) London, HMSO (1973) p. xxii
26. ibid.
27. G. Stigler and D. Blank *The Demand and Supply of Scientific Personnel* New York, National Bureau of Economic Research (1957) pp. 57ff
28. J. E. Tilton *International Diffusion of Technology: the Case of Semiconductors* Washington, Brookings Institution (1971) p. 34
29. Z. Griliches 'Research Costs and Social Returns: Hybrid Corn and Related Innovations' *Journal of Political Economy* (1958)
30. B. Weisbrod op. cit.
31. P. M. S. Jones 'The Use of Cost–Benefit Analysis as an Aid to Allocating Government Resources for Research and Development' in J. N. Wolfe (ed.) *Cost Benefit and Cost Effectiveness* London, Allen & Unwin (1973) p. 173
32. K. Grossfield and J. B. Heath 'The Benefit and Cost of Government Support for Research and Development: a Case Study' *Economic Journal* (1966)
33. ibid.
34. R. L. R. Nicholson 'The Practical Applications of Cost Benefit Analysis to Research and Development Investment Decisions' *Public Finance* (1971)

THE GOVERNMENT'S ROLE 155

35. M. Peck and F. M. Scherer op. cit. chapter II
36. Public Accounts Committee, *Sixth* and *Seventh Reports* (HC 335 and 353, Session 1972/73) op. cit.
37. M. Peck and F. M. Scherer op. cit. p. 19
38. NRDC *Annual Report and Accounts, 1953/54* (HC 27, Session 1954/55) London, HMSO (1954) p. 2
39. NRDC *Annual Report and Accounts, 1966/67* (HC 606, Session 1966/67) London, HMSO (1967) p. 6
40. NRDC *Annual Report and Accounts, 1970/71* (HC 553, Session 1970/71) London, HMSO (1971)
41. Monopolies Commission *A Report on the Boots–Glaxo Beecham Proposed Mergers* (HC 341, Session 1971/72) London, HMSO (1972) p. 10
42. NRDC *Annual Report and Accounts, 1966/67* pp. 50–4
43. Select Committee on Science and Technology *Third Report* (HC 420, Session 1972/73) London, HMSO (1973)
44. ibid. p. xliii
45. ibid. p. xviii
46. K. Grossfield 'Inventions as Business' *Economic Journal* (1962)

7 Technological Change in Industry

This chapter is devoted to a study of some important facets of the process of technological change in industry. Two aspects of this process have already been examined in detail: in chapters 3 and 4 the relationship between absolute and market size and invention and innovation were discussed. In this chapter we shall be concentrating on two main issues: the distribution of R & D expenditure in industry and the determinants of that distribution; and the diffusion of innovation.

7.1 R & D expenditure in industry

7.1.1 *Distribution of R & D between industries*

We have already pointed out some of the limitations of official R & D statistics (p. 51), in particular that they exclude 'informal' R & D work, which may be especially important in the case of the small firm. However these statistics are the best currently available and it is unlikely that the greatly increased cost of obtaining greater accuracy could be justified.

Columns (2) and (3) in table 7.1 provide a summary picture of the distribution of R & D expenditure in manufacturing industry in the UK in 1968–69 (the industries are ranked in terms of column (8)). R & D in manufacturing industry accounts for approximately 98 per cent of all industrial R & D, the rest being undertaken mainly in construction. The figures in column (2) relate to *expenditure* by private industry, nationalised industries and cooperative research associations in each of the product groups; they give no indication of the *sources of finance*. Column (4) remedies this omission to some extent by giving the percentage of industrial expenditure financed

TECHNOLOGICAL CHANGE IN INDUSTRY   157

Table 7.1
R & D expenditure in UK manufacturing industry

(1) Product Group(s)	(2) Total R & D expenditure 1968–69 (£m.)	(3) % of Total	(4) % of total R & D expenditure in each industry financed by govt	(5) % of total government-financed R & D expenditure in industry	(6) Privately financed R & D expenditure (£m.)	(7) Net output 1968 (£m.)	(8) (6) as % of (7)
Petroleum products	12.0	1.8	0.6	—	12.0	135.3	8.9
Electrical engineering	172.3	26.3	29.3	24.2	121.9	1375.3	8.9
Aerospace	166.7	25.4	83.1	66.6	28.1	478.2	5.9
Pharmaceutical products plastics, chemical and coal products	83.8	12.8	3.4	1.8	48.8	2761.0	5.9
Motor vehicles	45.1	6.9	3.5	—	43.6	1042.8	4.2
Scientific instruments	15.9	2.4	20.5	1.6	12.6	299.3	4.2
Railway equipment	2.1	0.3	—	—	2.1	59.7	3.8
Mechanical engineering	63.7	9.7	13.2	4.0	55.2	1963.0	2.8
Rubber and rubber products	6.3	1.0	0.5	—	6.2	280.3	2.4
Non-ferrous metals	6.7	1.0	5.2	—	6.4	575.0	2.3
Stone and clay products	12.0	1.8	2.7	—	11.6	358.6	2.0
Iron and steel	11.1	1.7	3.6	—	10.7	788.7	1.4
Textiles and man-made fibres	13.3	2.0	3.0	1.8	20.4	1841.4	1.1
Food, drink and tobacco	20.6	3.1	1.1	—	12.9	1058.3	1.2
Metal products, n.e.s.	7.9	1.2	—	—	7.6	921.4	0.8
Ships and marine engineering	3.2	0.5	34.7	—	2.1	273.6	0.8
Other manufactures	2.8	0.4	—	—	2.8	350.6	0.8
Timber, furniture, paper, etc	8.6	1.3	3.4	—	8.3	1653.5	0.5
Clothing, footwear, leather	1.4	0.2	6.7	—	1.3	555.8	0.2
All manufacturing	656.4	100.0	31.7	100.0	448.4	15,289.4	2.9

Sources: *Research and Development Expenditure* (Central Statistical Office) London, HMSO (1973) table 16C; *Report on the Census of Production* (1968) (No. 156) London, HMSO (1972) table 1

by government. The remainder is nearly all privately-financed R & D (column (6)).

Three points of particular interest emerge from a study of these columns. Firstly, industrial R & D expenditure is very heavily concentrated in a few industries. Aerospace, electrical and mechanical engineering and chemicals and allied products account for three-quarters of industry's total expenditure (compared with their 43 per cent share of net output). Secondly, industries vary widely in their dependence on government finance. Most of the product groups mentioned obtain less than 10 per cent from this source, but the shipbuilding and marine engineering, electrical engineering and aerospace industries receive much more. In the last-mentioned industry, 83 per cent of R & D is financed by government. Finally, as mentioned in the last chapter, government-financed industrial R & D is heavily concentrated in aerospace and electrical engineering. The effects of government spending in industry have already been discussed. The relative importance of different industries and of government finance in industrial R & D expenditure in the US is very similar to that for the UK outlined above[1] (but see p. 160 below).

7.1.2 *Research intensities*

Comparison of absolute R & D expenditures by different industries is likely to be of little value in economic analysis, unless these expenditures can be related to some measure of the industry's size. Measures of 'research intensity' are therefore necessary (p. 54).

The research intensity figures used in column (8) in table 7.1 relate to privately financed R & D expenditure-to-net output ratios. We should note the following points about the data. Firstly, the R & D figures are for 1968–69 whereas 1968 net ouput figures are used. Thus the intensities are intended to give an indication of the *relative* position only of different industries. Secondly, the R & D figures were collected on a product group basis, with firms allocating their total R & D expenditure between different groups where necessary, whereas the net outut data are on an establishment basis: an establishment is allocated to a particular industry if the products of that industry account for the biggest single share of its output. This difference however is unlikely to affect substantially the magnitudes involved. Thirdly, the ratios refer to *privately financed* R & D. Government finance has been deducted from the totals. This has been done because it will usually be the level of privately financed R & D that economists will be trying to explain. However, as pointed out in the previous chapter (p. 132), this expenditure may not truly reflect the amount of R & D that might *otherwise* be undertaken in the absence of government funds because the latter may itself influence how much industry is prepared to spend. For example, government money may induce an industrial recipient to spend less from his own budget. Conversely, it may act as a pump primer, inducing firms to spend more on their privately financed work. Because government R & D expenditure in industry may lead industry to curtail its privately financed programme, Tilton has suggested that for some purposes of analysis, *total* industrial R & D data should be used since they may come nearer to what industry itself would finance if no government funds were available.[2]

Finally, the date refer only to R & D performed and output produced in the UK. Clearly the picture is complicated by the existence of multinational corporations. (These are particularly important in science-based industries.) The location of the R & D undertaken by these corporations may bear little relationship to the location of their output, with the result that the *UK* research intensity figure for these firms may have little meaning. Similar problems are likely to exist at the industry level, although their extent is not known.

Despite these qualifications, it is likely that the figures in column (8) of table 7.1 give a fairly good indication of the importance of R & D in different industries. It is apparent that industries differ widely in their commitment to this activity. There are also considerable differences within industry groupings that are not apparent from the aggregate figures given in the table. For example, in 1960 the FBI (now the CBI) provided more detailed data on research intensities (measured by the ratio of QSEs to total employment). Their survey, which was limited to FBI members only, showed that the research intensity of the man-made fibres sector of the textiles industry was almost as high as that for electrical engineering.[3]

The ranking of US industries by research intensities would provide a very similar order to that found in table 7.1, although in almost all industries intensities are greater in the United States. Freeman for example has shown that in 1958 the ratio of US research intensity to UK intensity (using *total* R & D expenditure-to-net output ratios and a research exchange rate) was 1.8.[4] In only three industries – aircraft, non-ferrous metals and textiles and apparel – was this ratio less than one. This difference in research intensities does at first sight appear to have important implications for UK–US competition : if the US spends relatively more on R & D, it may be expected that it will have a higher innovative record with both its products and processes and therefore be more competitive. However this conclusion cannot be reached quite so easily, since the US government spends a greater share of its R & D budget in industry : this in turn may boost R & D intensities. It may also be argued that US R & D productivity is lower and that industry is spending more than is optimal on R & D. However there is no evidence to support either of these views; Freeman also points out that the effect of greater US government spending in industry is unlikely to affect fundamentally the original picture. It is probably true that the higher US research intensities are at least partly a result of different management attitudes to R & D and to the fact that more US firms are above the threshold level.

What accounts for the differences in R & D commitment *between* industries? One hypothesis that seems reasonable is that the *profitability* of R & D in different industries varies. Certainly, at the micro level a few studies have shown the significance of profitability as one important determinant of a firm's R & D budget (see below). The profitability of R & D is likely to vary across industries for a number of reasons. Firstly, 'technological opportunity' is likely to differ : a given volume of R & D in one industry may produce much bigger results, eg in cost saving terms, than in another. This may be due in part to differences in the extent to which the underlying science is developed. Most of the 'science-based' industries for example are supported by considerable research programmes in universities and elsewhere.

Secondly, the nature of the demand facing different industries varies. In some industries it is highly sensitive to technological change : only the *latest* product will sell. This attitude is particularly important for example in the drugs industry where the fortunes of individual companies depend heavily on the extent to which their product is a leader in its field (see p. 89). In the more traditional industries, being at the forefront of technological development may not be so important in sales terms.

Thirdly, differences in industry growth rates may lead to differences in R & D profitability : a rapidly growing industry may make invention and innovation more attractive because the introduction of new products is less likely to upset existing market patterns, and therefore to invoke retaliatory action, as it might do in a declining or static market. Rapid growth will also provide greater opportunity for recouping R & D investment costs, as well as for increasing the supply of funds (growth is usually linked with higher profitability). The above discussion assumes that causality runs from growth to R & D. However it may also be the other way : demand may grow as the result of an innovation.

The relationship between output growth and research intensities is a complex one and it is worth pausing at this stage to look at the empirical evidence. Freeman showed that in seventeen UK industries growth of output between 1935 and 1958 was highly correlated with total research intensity in 1958 ($r=0.95$). For US industry the results also showed a strong positive association ($r=0.74$). Freeman is careful however not to commit himself to a causal relationship : '. . . the figures may only show that relatively rapid growth and high research expenditure are joint characteristics of new industries'.[5] Leonard has looked at the relationship between private research intensity and sales growth for sixteen industrial groups in the United States covering 97 per cent of all manufacturing sales, for several different dates and periods in the 1950s and 1960s.[6] Sales were lagged behind R & D expenditure so that any growth or decline would be measured beginning with the second year after the R & D input. His results seem to support the hypothesis that causality runs from R & D to sales; for example $r=0.8$ when 1967 research intensities are correlated with 1958–67 sales growth. He also found that reverse causality was not sustained by the data. Tilton however has criticised Leonard on the grounds that he did not use the most meaningful R & D measure. Tilton shows that, when government-financed R & D is included in the research intensity measure, the association found by Leonard becomes much weaker.[7] This inclusion of government finance is justified on the grounds mentioned earlier that privately financed R & D may be *reduced* when government contracts are provided. The empirical data on the relationship between research intensity and output growth have not therefore yielded unambiguous conclusions and it is likely that causality runs in both directions.

The fourth factor that may influence the profitability of R & D (in its widest sense) may be the absolute size of firms and market structure of the industry. There are a number of issues involved here. Where firms are small in absolute terms, viable R & D programmes may become prohibitively expensive (chapter 3). As far as market structure is concerned, an oligopolistic industry may give rise to extensive duplication in research thereby raising the overall research intensity. On the other hand, monopoly power may cause firms to reduce their R & D commitment (chapter 4).

Fifthly, some industries may find it more economic to 'buy in' R & D through purchases of capital goods and materials from other industries rather than to engage directly in R & D itself. Some industries may have greater access to the results of work undertaken outside industry, eg in government research laboratories.

Finally, a diversified industry may make R & D more profitable because there will be greater opportunity for the surprise results of R & D to be utilised. Minasian has also made the point that such surprise results may be more readily *recognised* and therefore utilised in a diversified industrial environment.[8] (The casual link between R & D and diversification may also operate in the reverse direction, with results from R & D leading to the addition of new product lines.)

There are sound reasons therefore for expecting variations in industrial research intensities. It must be remembered however that some industries may suffer from incomplete or distorted information on the returns obtainable from R & D (see p. 127). It might be therefore that more expenditure on R & D might profitably be made, but it is not forthcoming because of poor information about the returns possible.

7.1.3 *R & D in the individual firm*

A further insight into the reasons behind differences in industrial research intensities may be gained by an examination of the determinants of R & D budgets in individual firms. In 1957 Carter and Williams categorised the R & D decisions of the firms they investigated into three groups.[9] In *fully considered* decisions, a firm attempted to specify its R & D needs in the light of its objectives and to relate this to the costs involved. In an *elliptical* decision, calculations started from the premise that there were one or two limiting factors (eg the size of the research buildings) that was critical to the decision, without any consideration being given to whether such obstacles could be overcome. *Topsy* decisions involved very little positive action : things were just allowed to carry on as before. Carter and Williams found that fully considered decisions were made in less than ten per cent of the companies involved in their survey. They were of course writing more than fifteen years ago; it is likely that companies have generally become more sophisticated in their decision-making procedures since then. Furthermore it must be remembered that even Topsy-type decisions may in fact reflect far more fundamental determinants than is apparent at first sight, and may result from a stability in these determinants over time.

Grabowski has constructed a model to explain differences in research intensities in firms in three industries (drugs, chemicals, and petroleum-refining) over the period 1959–62.[10] Three independent variables were incorporated into his model. The first was an index of the firm's research productivity over a previous period. The underlying reasoning behind the inclusion of this variable was that a firm's estimate of future returns on existing projects will depend to some extent on past results, which are likely to have varied from firm to firm. The higher the productivity of past research, the higher the commitment to current research is likely to be. Research productivity however was measured by the number of patents granted in a particular previous period divided by the number of scientists and engineers employed during the period when these patents were originally conceived. These patent data are subject to well-known limitations which should now be clear to the reader (see p. 33). They are nevertheless attractive because of their availability and because they have at least to satisfy a roughly consistent minimum standard.

The second variable was an index of diversification. Grabowski included this variable on the grounds that the more diversified a firm is, the higher its profit expectations from R & D are likely to be because it has a wider 'net' to catch unexpected research results.* Finally, an index of a firm's internally generated funds

---

* The diversification measure used is the number of separate five-digit SIC product classifications in which a firm produces. This measure is also subject to considerable limitations although it is certainly a start. See M. Gort *Diversification and Integration in American Industry* Princeton, Princeton UP (1962) pp. 9–11.a

deflated by sales was incorporated into the model on the grounds that the availability of internal funds is likely to play a significant part in the allocation of funds to R & D.

More formally, Grabowski's model may be expressed as follows:

$$\frac{R_{i,t}}{S_{i,t}} = b_0 + b_1 P_i + b_2 D_i + b_3 \frac{I_{i,t-1}}{S_{i,t}}$$

where $R_{i,t}$ is the R & D expenditure of the ith firm in the tth period; $S_{i,t}$ is the sales revenue of the ith firm in the tth period; $I_{i,t-1}$ is the sum of after tax profits plus depreciation and depletion expenses in the $t-1$ period; P_i is the research productivity measure referred to earlier; and D_i is an index of diversification.

For the chemicals and drugs industries the explanatory power of the model is quite good (with $r^2=0.63$ and 0.86 respectively, with all the coefficients being significant at the one per cent level). However for firms in petroleum-refining the results are much less satisfactory ($r^2=0.29$, b_2 not significant at the one per cent level). Grabowski's explanation for the last mentioned result is based mainly on two factors. Firstly, patents are likely to be a poor measure of R & D output in this process industry 'because firms will often wish to keep knowledge of such inventive activity concealed from their competitors'. The reasoning behind this claim however is not given and is not immediately clear. Secondly, R & D in this industry is somewhat less important in competitive terms and therefore more subject to fluctuations in other activities.

Grabowski's results are interesting in that they suggest that R & D budgets are determined as a result of economic factors and not simply on a completely *ad hoc* basis. Furthermore, they also suggest that R & D expenditures may be manipulated by altering, for example, the tax treatment of companies. This may be an alternative policy to the direct provision of government funds. It should be noted however that Grabowski's results are limited to industries that are likely to be very sophisticated in their approach to R & D spending. His model might not perform so well in industries where research management is of a much cruder form.

Mansfield has also given a prominent place to the profitability of a firm's (private) R & D in the determination of budgets for the latter.[11] His basic model is as follows:

$$r_i(t) = R_i(t-1) + \theta_i(t)[R_i(t) - R_i(t-1)]$$

where $r_i(t)$ is the ith firm's R & D budget for year t.; $R_i(t-1)$ is actual R & D expenditure in year $(t-1)$; $R_i(t)$ is the desired R & D expenditure for year (t); $\theta_i(t)$ is the fraction of the way the ith firm moves towards the desired R & D expenditure; and $R_i(t)$ is assumed to be greater than $R_i(t-1)$ although this is not crucial to the model.

It can be seen that in this adjustment model Mansfield explicitly includes the possibility that a firm may not reach its desired R & D budget in a given year because of expansion costs and uncertainties associated with increasing the R & D establishment beyond the previous year's size. $R_i(t)$ is determined essentially by the expected profitability of R & D projects proposed: all projects with an anticipated rate of return above a certain minimum would be accepted. The resulting desired level of R & D spending would be greater for larger firms than for small firms, because the former would tend to generate more proposals for R & D (for example because higher sales volume of the larger firm may lower the unit cost of R & D and thereby make certain proposals more attractive). The extent to which the previous year's budget is expanded, $\theta(t)$, is assumed to be directly related to 1) the magnitude of the gap between $R_i(t)$ and $R_i(t-1)$, (a small increase is likely to cause fewer problems than a large one), and 2) the ratio of the firm's profits to its actual R & D expenditure in the previous year. This second assumption is based on the idea that finance for R & D expansion will have to come from profits. The greater the amount of profits that would be absorbed in an increase, the more cautious a firm is likely to be. Mansfield tested his model using data from eight firms in chemicals and petroleum and found that his model explained R & D budgets in these firms reasonably well. While the data used are very crude and limited,* Mansfield's work provides further confirmation for the idea that R & D expenditures, like other outlays of a firm, are strongly influenced by economic considerations. Schmookler's work (see pp. 35–8) of course add further support for this hypothesis.

* One limitation that may be particularly important in interpreting the results is that only about half of the firms Mansfield approached would give him the data he needed. This may imply that these firms were more likely to conform to the model. The non-co-operating firms may have refused to help precisely on the grounds that their R & D budgets were determined by haphazard and non-economic considerations.

In addition, Mansfield has also shown that expected profitability also plays an important part in the *internal* allocation of funds *between* R & D projects.[12] In one large company he found that about fifty per cent of the distribution of funds between projects was explained by profit expectations. This does of course still leave a substantial unexplained residual. Other determinants are obviously important and may be related to non-economic variables. For example, research workers may attempt to orientate their work towards projects that are most likely to meet with the approval of the scientific community; such projects may not however be of great immediate commercial significance. In interpreting Mansfield's work it must be remembered that he focuses on *expected* profitability as a determinant of R & D expenditures. The *accuracy* of such predictions may however be very low (see p. 20).

7.1.4. *R & D and profitability*

So far most of our attention has been centred on expected profitability as a determinant of R & D budgets and programmes. There are however two further aspects of the profitability–R & D relationship that also require attention. Firstly, past profitability may *determine* R & D budgets, as the former may act as a budget constraint on the latter. Allowance for the role of profits as a source of finance is of course made explicit in both Grabowski's and Mansfield's models. Or the position may be simpler: firms may regard R & D as something of a 'luxury' activity which can only be afforded in times of affluence, ie high profitability. This latter possibility however seems unlikely to exercise a *major* influence on R & D budgets, although it is likely that in times of high profitability firms will be more willing to finance more speculative projects and to give their research staff more leeway in times of financial strength. Secondly, the causation may be reversed, with high R & D expenditures leading to higher overall profitability. Minasian has looked at these two hypotheses in his study of R & D expenditure in chemicals and found that research and development at the beginning of the period studied explains end-period profitability better than vice-versa.[13]

7.1.5 *The nature of industrial R & D*

As far as the kind of R & D carried out in industry is concerned, there is evidence on both sides of the Atlantic that the majority of such work is directed towards improving existing products and processes rather than developing new technologies. In America a McGraw-Hill survey revealed that, in 1958, 50 per cent of industrial R & D was on product and process improvements. The survey also showed that 91 per cent of firms required a pay-off from their R & D within five years.[14] Many inventions will require much longer than this to come to commercial fruition (see below). In Britain, an FBI investigation showed that in 1960 over 50 per cent of R & D in industry was devoted to 'major and minor improvements' and to technical services.[15] ICI, which spent £29m. on R & D in Britain in 1969, allocated 60 per cent of its budget to the *improvement* of products and processes.[16]

In interpreting these figures, it must be remembered that the process of technological advance probably consists of a constant stream of what, taken individually, are relatively minor improvements and adjustments, rather than of a few big 'jumps'. Furthermore, the element of chance in R & D is still important: major new technologies may still emerge from work that was intended only as 'improvement' R & D. Nevertheless, it is likely that many major inventions and advances will come from outside industrial laboratories, because of the type of work the latter usually undertakes.

7.1.6 *Distribution of R & D expenditure between basic research, applied research and development*

In table 7.2 a breakdown of total (current) R & D expenditure in UK industry by type of activity is given. There are considerable definitional problems in a breakdown of this kind, although once again, provided the data are treated carefully, they may be useful in giving some idea of the distribution of resources between types of work.

Manufacturing industry as a whole spends only 4 per cent on basic research. This figure and those for applied research (22 per cent) and development (74 per cent) are very similar to those for US industry.[17] The small allocation to basic research is as would be

Table 7.2
Current expenditure on R & D performed within UK manufacturing industry (% of total in each industry)*

Product group	Basic Research	Applied Research	Development
Electrical engineering	2.3	16.6	81.1
Aerospace	2.0	9.0	88.6
Chemicals and allied products	7.9	43.2	48.9
Motor vehicles	1.2	11.4	87.4
Scientific instruments	4.7	17.8	77.6
Railway equipment	—	62.7	37.3
Mechancial engineering	8.1	24.2	67.7
Rubber and rubber products	0.5	41.5	58.1
Non-ferrous metals	4.9	43.9	51.2
Stone and clay products	10.5	30.8	58.7
Iron and steel	7.3	53.0	39.8
Textiles and man-made fibres	3.3	38.6	58.1
Food, drink and tobacco	8.4	40.8	50.8
Metal products n.e.s	6.2	23.4	70.4
Ships and marine engineering	3.4	13.9	82.7
Other manufactures	3.1	19.9	77.1
Timber, furniture, paper printing, publishing	4.7	30.2	65.1
Clothing, footwear, leather	8.0	42.3	49.7
All manufactures	4.0	21.7	74.3

* Includes expenditure by public corporations and research associations.

Source: *Research and Development Expenditure* London, HMSO (1973) table 18D

expected from the character of the work and from the shortness of the period in which firms expect a return for their R & D (see above).

There are wide variations in the extent to which individual industries support basic research, although in no case is it over 11 per cent of the total. No simple explanation for this variation exists, although some tentative reasons may be advanced. Firstly, it might be expected that industries with a greater appreciation of the value of R & D generally might spend more on this type of research. (It may also be true that such industries will have a greater appreciation of what constitutes basic research, and will therefore be more discriminating in allocating expenditure to this heading when filing their statistical returns. More traditional industries may not be so discriminating; this may partly explain the apparently high commitment of many of these industries to basic research.) However there is no clear relationship between research intensity (one possible measure of such appreciation) and the percentage spent on basic research. Secondly, industries in which the average size of firms is large might spend relatively more, because for them the economic returns from basic research findings may be greater; again however this does not appear to be a satisfactory explanation. Several industries, eg clothing, leather and footwear, have relatively big commitments to basic research yet their average size of firms is small. Furthermore, US evidence suggests that large firms do not spend more in this field than smaller firms. Indeed, as far as federally financed industrial R & D is concerned, they spend less.[18] Thirdly, industries vary in the extent to which external bodies, eg in government and higher education, undertake basic research; this will clearly influence their own commitment in this field. Fourthly, the costs of other activities in the R & D spectrum are likely to vary from industry to industry. In some, eg heavy electrical engineering and aircraft, expensive testing equipment and pilot plant may be necessary for development work, whereas in others capital costs even in development may be relatively low. This will influence the percentage of R & D spent on basic research. Finally, basic research in some areas might be more economically productive than in others.

It should be noted that in absolute terms, industrial basic research is concentrated in a few industries: chemicals and engineering accounted for 60 per cent of all such work in 1968–69.

7.2 Patents and royalties

Patents accepted by the UK Patent Office during 1960 are shown in table 7.3, classified by industry. It should be remembered that the industry to which a patent is finally classified depends on the subject matter of the patent and *not* on the industry in which the patent originated. By and large, it is probably true to say that most industrial inventions will originate in the industry to which they are classified, but there are important exceptions where inventions have occurred outside the traditional boundaries of an industry. A second point to note is that a substantial proportion of the patents listed in the table does not come from within *UK* manufacturing industry itself. According to Taylor and Silberston about 75 per cent of all UK acceptances in 1960 were for patents that came either from corporations or individuals overseas, or from UK

Table 7.3
Patents and royalties in British manufacturing industry

Industry	(1) New Patents 1960*	(2) % of total	(3) New patents 1960, per £10,000 of R & D expenditure 1958	(4) Royalties received 1968 (000s)	(5) Royalties paid, 1968 (000s)	(6) (4) as % of net output	(7) (5) as % of private R & D expenditure 1968/69	(8) Ratio of (4) to (5)
Wood and paper products	1190	3	6.0	3312	6988	0.42	84.1	0.47
Bricks, pottery, glass	1180	3	3.7	5805	2421	0.42	20.8	2.40
Other manufacturing	1330	3	3.2	3430	7169	0.22	43.1	0.48
Textile, man-made fibres Clothing, leather, footwear	2310	6	2.7	8041	4999	0.30	55.5	1.61
Metal manufacture	1860	4	2.3	679	1458	0.08	8.5	0.47
Chemicals and petroleum products	9000	21	2.1	21,015	19,585	0.67	32.2	1.07
Engineering and scientific instruments	21,890	52	2.1	14,021	50,955	0.38	26.8	0.28
Ships and marine engineering	300	0.5	1.3	734	1002	0.36	47.7	0.73
Drink, food tobacco	520	1	0.9	4043	3667	0.19	17.9	1.10
Aircraft, motor vehicles	2390	6	0.2	5374	5000	0.32	6.7	1.07
Railway equipment								
All manufactures	41,970	100	1.4	66,454	103,244	0.43	23.02	0.64

* British patent acceptances from both UK and foreign sources.

Sources: As table 7.1, and (first three columns) C. Taylor and Z. Silberston *The Economic Impact of Patents* Cambridge, Cambridge University Press (1973) table 4.3

independent inventors.[19] This percentage is likely to vary considerably from industry to industry.

The table shows a very heavy concentration of patent acceptances in the engineering and chemical field. These industries accounted for 74 per cent of all patents, compared with 53 per cent of all privately financed R & D in 1968–69 (table 7.1). On the other hand the corresponding figures for aerospace, railway equipment and motor vehicles (taken together) were 6 and 16 per cent.

Column (3) in the table looks more closely at the relationship between patents and R & D expenditure. The expenditure relates to 1958; thus some allowance is made for the lag between expenditure and patent acceptance (p. 40). Little weight can be attached to these figures because of the data limitations described earlier. Neverthless, it is hard to escape the very broad conclusion that ability and/or propensity to patent varies considerably across industries. Furthermore it is not related to the research intensities of the industries concerned. This variation in industrial patenting is not unexpected and it probably arises for two main reasons. Firstly, the patentability of R & D output is unlikely to be constant across industries. It is also likely to vary with firm size : larger firms are more likely to engage in the kind of R & D work whose output is relatively less patentable than are smaller firms. This latter fact will affect the figures because of differences in industrial structure between industries. Secondly, even where patenting is possible, industries are likely to vary in their response to this opportunity. Market structure and the degree to which secrecy can be maintained are likely to be important influences here.

7.2.1 *Royalties*

Columns (4) and (5) of table 7.3 give details on royalties received and paid for in 1968. These data have been taken from the Census of Production and, as with the statistical material discussed earlier, have several characteristics that might limit their usefulness. Two of these characteristics are of particular interest here. Firstly, the figures cover not only receipts and payments relating to patents but also transactions relating to trademarks 'and/or any lump sums payable or receivable on revenue account'. Thus the figures are somewhat broader than those required for our discussion here, although with a few exceptions royalty transactions will probably account for the bulk of all receipts and payments. Secondly, the figures were collected on a *product* basis. This means that they may not correspond exactly with the net output figures. Column (6) gives some idea of the importance of royalty receipts in terms of the industry's size. It is apparent that, although varying from industry to industry, they never account for more than one per cent of net output.

Column (7) relates royalty payments to R & D expenditure in 1968–69 – an extremely crude measure of the extent to which a firm 'buys in' its technology. (One limitation of the figures is that the R & D expenditure data relate to *current* programmes, while royalty payments stem from R & D undertaken sometime in the past). Buying in technology may be an economically rational policy to undertake in certain circumstances, although only very rarely is no adaption of any kind likely to take place. In extreme cases, given the superiority of a competitor's technology, buying in may be the only alternative to not entering a field or even to bankruptcy. Overseas producers of flat glass for example had to obtain licences from Pilkington for their float process if they were to compete effectively. For some industries this dependence seems very high. Royalty payments are of course usually affected by the *scale* of operations (as are royalty receipts). The higher the output, the bigger in absolute terms the total royalty payable. Thus the bare figures in the table may give a misleading impression of a firm's overall dependence on external sources of technology: a relatively minor invention may yield very large royalties if the scale of output is high. R & D expenditures however are largely unaffected by scale of output.

The final column gives the ratio of receipts to payments. Manufacturing industry as a whole is in substantial deficit. The highest surplus is found in bricks, pottery and glass. This is almost entirely due to the foreign revenues from flat glass. In 1963, before this innovation had been fully developed, the ratio was only 0.18, an indication of how the picture may change drastically as the result of one major breakthrough.

In interpreting the ratios in column (8), it is worth bearing in mind some of the findings made by Taylor and Silberston in their study of patents and licensing. Firstly, nearly all the licence agreements they came across were with other firms. Thus an industry is unlikely to be in surplus or deficit as the result of transactions with government or independent persons or bodies in the UK. Secondly, over 75 per cent of these licence agreements were with firms who were operating in the same fields.[20] It follows from this that the surplus or deficit is unlikely to be the result of UK industries trading know-how *with each other*. (This is of course borne out by the overall deficit for manufacturing.) From these two points it follows that the ratios provide a rough indication of the extent of foreign trade in patents, know-how etc in different industries. This conclusion is supported by Taylor and Silberston, who found that the preponderance of licensing by British firms is with foreign countries. Some of this international flow of technological payments will be between parent and branch companies. The size of these flows may be determined not solely on economic criteria but on other grounds. For example, tax laws may encourage a foreign parent to take its profit from an overseas subsidiary in the form of inflated licence payments.

It must be stressed again that a deficit does not necessarily reflect adversely on the industries concerned: 'buying in' technology may be an optimal policy. Furthermore, low royalty receipts may be a direct result of the exploitation of a *strong* patent position with the patent-holder refusing to license other companies, producing the goods itself and selling at monopoly prices both at home and abroad. On the other hand, if Taylor and Silberston's findings that UK firms do not usually license domestic competitors applies in other countries too, then it follows that British firms who obtain licences are unable to operate in the markets in which the invention was first introduced. This may limit their growth potential.

7.3 Diffusion of innovation

In this section we shall examine some of the factors affecting the pace at which inventions and innovations are accepted and taken up by industry for commercial use. We first look briefly at the lag between invention and innovation and then at the diffusion of innovation.

7.3.1 *The lag between invention and innovation*

A distinction was made in chapter 1 between invention and innovation, although it was pointed out that this distinction is often a fuzzy one. However several writers have attempted to measure the interval between invention and innovation. Enos for example, drawing on other work, has provided data on this lag for thirty-five inventions over the last two and a half centuries.[21] The invention data are for '. . . the earliest conception of the product in substantially its commercial form . . .'; the innovation data are for the '. . . first commercial application or sale . . .'. Even given these definitions however, there is considerable room for differences in interpretation. The sample gives a lag with an arithmetic mean of 13.6 years which is reduced to 11.6 years if the fluorescent lamp which had the particularly long lag of seventy-nine years, is excluded (see pp. 81–2). Mechanical inventions tend to require the smallest time interval, with pharmaceutical and chemical inventions next. The inventions with the longest lag were those in the electronics field. Enos found that 'The interval appears shorter when the inventor himself attempts to innovate, than when he is content merely to reveal the general concept'. It is probably true to say that an inventor who is intimately involved with the commercial application of his work is likely to achieve quicker results than if his contact with the innovating corporation is only at arm's length. However, Enos does not ask whether the timing of the innovation was *optimal;* hovercraft development in the UK (see part 3) provides examples that indicate only too clearly the painful consequences of too-early marketing. It may be that the inventor who is also closely connected with the innovating process is more prone to make this mistake.

The time lag will be determined by a host of factors, including the extent of the investment required; the market structure of the industry concerned; the certainty of market prospects; the extent of the advantages that the new process or product has over existing products or processes; and the degree of government involvement. These issues are examined more fully later when the diffusion of innovation is discussed.

What is interesting about Enos's figures – a fact pointed out by Blair[22] – is that the average time lag is equivalent to at least two-thirds of a patent's life. This suggests, *prima facie,* that patents are unlikely to provide a very effective form of protection. It is also worth noting that there appears to have been hardly any reduction in this lag over time.[23] This evidence seems however to conflict with that produced by Lynn.[24] This kind of conflict is probably inevitable given the wide scope for interpretation possible in the dating of inventions and innovations. We now turn to a more general examination of the diffusion of innovation among firms.

7.3.2 Diffusion of innovations among firms: some theoretical considerations

We have already made a distinction (pp. 17–8) between product and process innovations although the two types are closely interlinked: one firm's product innovation will often be incorporated into another's process; or a new product may imply a new process. However, for ease of exposition the basic distinction is maintained. Only process innovation will be considered in detail here although the framework outlined below can be adapted to deal with product innovations. We shall consider the simple case of an innovation which, via a new piece of equipment or machinery, enables a given good to be produced more cheaply than hitherto.

The rate of diffusion is concerned with the speed at which a given innovation is taken up in industry. We might measure it in terms of the percentage number of firms in an industry using the innovation or in terms of the percentage of that industry's output that it is used to produce, at different points in time. For most purposes the second measure will be the most useful although it will often present greater problems of data collection. Both methods also raise the difficult problem of the relevant base to use in the calculation of these percentages. Industries, even at the level of Minimum List Heading, usually produce a very diverse output. Process innovations will often be appropriate for only part of an industry's production. Further difficulties may arise if the applicability of an innovation increases over time as the result of technical improvements.* Thus, the measurement of diffusion is bound to be based, to some extent at least, on arbitrary judgement.

Before examining the factors affecting diffusion rates in detail, let us take a 'snapshot' view of the range of techniques likely to be in use in an industry in a given point in time. A number of

---

* Nabseth gives examples of both difficulties. The limited applicability of an innovation is well illustrated by the tunnel kiln which can only be used for certain grades of clay. Increasing applicability over time can be illustrated by the basic oxygen process. Initially this process 'could not be adapted for big plants, for producing specialised steels, or for processing high-phosphoric ores'. See L. Nabseth and G. F. Ray *The Diffusion of New Industrial Processes* Cambridge: Cambridge UP (1973) p. 297.

empirical studies have shown that there are often wide differences in efficiency (measured in various ways) between plants and firms in the same industry.[25]

To what extent are these differences explained by variations in the techniques employed, or are there other important explanatory factors? Are there sound economic reasons for these differences?

One of the most detailed studies of these issues has been made by Salter.[26] He constructed a simple model based on the following assumptions (these assumptions were made principally for expository purposes).

1) New techniques require new specialised capital equipment.
2) A plant consists of an integrated and indivisible fixed set of techniques. No piecemeal changes of individual pieces of equipment or machinery within the plant can take place.
3) All plants are homogenous apart from differences in techniques. For example, there are no differences in managerial efficiency or in the quality of labour.
4) Factor and product markets are perfectly competitive. Among other things this must mean that firms are perfectly informed about available techniques.
5) All plants work at 'normal' capacity.

In this theoretical world, the plants in a given industry may be represented as in figure 7.1. Operating costs per unit of output are measured on the vertical axis and output on the horizontal axis. O_n is the output capacity of plants built in period n, and AC is their unit operating cost. These plants are the newest and incorporate 'best practice' techniques. Similarly the plants built in period $n-1$ have unit operating costs of AE. The oldest plant produces output O_{n-t} at a unit cost of just below AD. Each block therefore represents total operating costs for each plant (or set of plants) built in a given period, operating at normal capacity.

Technical progress over time (via process innovations in the industry is illustrated by falling unit operating costs. This fall is made possible by the innovation and the fact that newer plants are likely to have lower repair and maintenance costs. In this model '... the plants in existence at any one time, are, in effect, a fossilised history of technology over the period spanned by their construction dates ...'.[27] It is possible of course that this fall in operating costs could be lessened or even reversed by the existence of a 'learning effect' in older plants, although there is some evidence to suggest

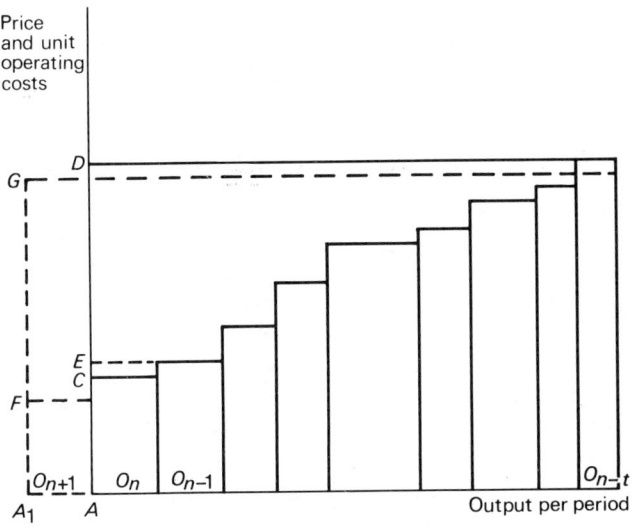

FIGURE 7.1

that this effect may not be of major significance in establishing a productivity differential between new and existing plants.[28]

On the basis of this model, we can now look in more detail at the economic factors affecting the diffusion process. If the diffusion of a process innovation is measured in terms of the percentage of an industry's output produced in plants incorporating the new technique, then it will be affected in three ways. Firstly, if old plant is scrapped, and not replaced, then existing plant incorporating a particular innovation will account for a higher percentage of the industry's output. Similarly this percentage will increase if, secondly, new plants incorporating the innovation are added to the industry's capacity without any scrapping occurring or, thirdly, if old plants are replaced by the new plants. In all three cases, the changes may occur either through existing firms expanding or contracting, or through the exit or entry of firms. In any event however, as the Salter model reminds us, the percentage is unlikely to reach 100 per cent because technical progress is a continuous process. A new plant embodying an innovation may be built but before this innovation has had time to permeate the whole industry, yet further innovation may be produced necessitating further change.

As fare as the scrapping decision is concerned, it will pay firms to abandon existing plants whose revenue is not sufficient to cover their operating costs and to provide a normal return on the present value of the site and working capital. These last two factors come into the picture because they often have alternative uses. Where a plant has a resale or scrap value, the plant's revenue would also have to cover the return obtainable on this amount. Depreciation calculations however are not relevant to the scrapping decision since, once 'free' capital has been converted into a physical form embodying a given technique, there is no possibility of reversing the process (except in so far as an individual firm may be able to resell or obtain a scrap value for his plant). Thus for the individual manufacturer, specific capital costs incurred sometime in the past should not enter into his decision on whether or not to scrap his plant. In short, 'bygones are bygones'. Costs resulting from physical wear and tear will of course be included in operating costs. (In the very short run even these costs may be ignored in the case of specialist equipment, provided revenue at least covers other operating costs.) The relevant factors to be taken into account when a new plant incorporating an innovation is *planned* are however different. In an *ex ante* evaluation, anticipated revenue would have to cover both operating *and* capital costs before it would be profitable to build such a plant. Thus while (past) capital costs are irrelevant to a scrap/operate decision relating to an existing plant, they must be taken into account when deciding whether to install a new plant, either as a replacememt for an old plant or as additional capacity. It will not pay a firm to replace an existing plant with a new one unless the former's operating costs exceed the capital and operating costs of the latter. As a new plant will usually last several years, some basis must be found for allocating its initial capital costs over the successive operating periods of the plant's life.

One way of doing this is based on the recognition that there are two costs associated with capital usage – interest and depreciation. Interest on the whole of the initial capital cost is payable in each operating period. A constant depreciation charge is levied in each period of the plant's life which, if invested at the company's cost of capital, will accumulate to the initial capital cost at the end of its life. Thus in each period a constant capital charge can be calculated. On the assumption that a plant has no scrap value, the annual capital charge A will be found by the following formula:

$$A = \frac{Cr(1+r)^n}{(1+r)^n - 1}$$

where C is the capital cost of the plant, r is the rate of interest and n is the plant's life.

There is one proviso to the replacement rule outlined above: in a case where a major technological breakthrough is expected it may pay a firm to postpone replacement in order to gain the advantages of later developments.

In a competitive system price will eventually fall to a level where only the operating and capital costs of the newest plant are covered. For example in figure 7.1 price will fall to D, where the latest plant (with output O_n) just covers its operating and capital costs (AC and CD respectively). Plant with output O_{n-t} will be the marginal plant, with price only very slightly above its operating costs. All other plants will be making some contribution to capital costs, since price is above operating costs. Now, say a new plant of capacity O_{n+t} with different techniques and operating costs of A_1F and capital costs of FG becomes available. Firms attracted by excess profits $(D-G)$ will build new plants, and price will eventually fall to G as output expands. But plant with ouput O_{n-t} will no longer covering its operating costs and will be scrapped.

The brief outline given above may help us in identifying some of the factors that are likely to be important at the micro level in the acceptance of new processes in a competitive industry. Operating costs of existing plants must be compared with the operating *and* capital cost of new plants. Comparison of operating costs alone is not enough. The weighting in favour of existing plant in such comparisons goes some way towards explaining why it may not be economic for firms to adopt best practice techniques for all their production. (This should always be remembered when interpreting diffusion data. The fact that a particular country or industry is slow in adopting a specific technique may not of itself indicate inefficiency: it may merely reflect the modern nature of its existing equipment.) Even with process innovations that effect major savings in operating costs, replacement of existing plant may not be worthwhile, especially if the latter has been built fairly recently and represents 'nearly best practice' techniques. Any change in the relative price of labour (wages will usually be a major component of operating costs) and capital will also affect the rate of adoption of innovation.

The picture is not so simple as painted however, for a number of reasons. One major problem arises on the empirical side. Although Salter drew on evidence relating to a few specific industries to show that the latest plants have higher levels of efficiency than existing plants, there is also other evidence which suggests that *in general* these plants do not necessarily have higher-than-average efficiency. This in turn implies that 'If new factories incorporate the latest equipment, then in most industries either new equipment is relatively unimportant as a determinant of productivity dispersion or there are systematic factors which . . . bias the labour productivity of new factories downwards'.[29] Another alternative, though less likely, explanation is that new factories do not usually embody best practice techniques.

Once outside the restrictive confines of Salter's model there are a number of other important factors that may influence the rate of diffusion. (Salter does in fact relax or at least discuss the assumptions on which his initial model is based.) Firstly, we have said nothing about uncertainty. Often a manufacturer does not know whether the performance of a new process or machine will be as claimed by those producing it. New machinery and equipment are frequently subject to extensive teething troubles, which may in turn affect operating costs. This problem clearly emerged from our study of hovercraft development (part 3). In some cases it may not be possible to specify the performance of machinery until a manufacturer has actually used it. In such cases many manufacturers may leave it alone until a number of others have tried it. Willingness to experiment with an innovation is likely to depend in part on how large an investment is required, and on whether it can be tested out on a small scale (see the following section).

Secondly, as Salter himself points out, technical progress does not usually come in units of plant size : it is normally of a piecemeal nature, affecting particular parts of a factory's production processes. The basic issues in scrapping, expansion or replacement decisions remain the same, but the position becomes much more complex because a new machine or piece of equipment in one area of production may upset the conditions of production elsewhere because of technical interdependence. For example, a new machine may require slightly different raw materials or produce slightly different output, both of which may necessitate adjustment elsewhere.

In some cases technical interdependence may mean that new processes are most profitable in situations where the rest of the equipment in a factory is also fairly modern. If this is so then it might be expected that, contrary to what would be expected from the Salter model, the plants with the older equipment and therefore lower productivity would have a slower rate of diffusion. (Eventually of course they may be forced to undertake drastic modernisation on a large scale or go out of business.)

Thirdly, and linked to the second point, considerable disruption costs may result from the installation of a new plant or machine. Factory layouts may have to be reorganised; the composition of the labour force may have to be modified, new management skills may be necessary and new production schedules established. Ray gives an example of such changes in his discussion of the introduction of numerically controlled machines.[30] Because the operator could no longer determine the pace of the machine, firms using this innovation had to switch from piece- to time-rate payments systems. Such disruption costs if they can be quantified would have to be included in the calculations. They might also lead to managers opting for a 'quiet life' by avoiding the introduction of new techniques. Their ability to do so will be determined in part by how competitive the industry is (see below).

Fourthly, if markets are not highly competitive then firms may not be forced to adopt the latest techniques even when they become profitable. Firms are usually to a greater or lesser degree insulated from the full blast of competition; they are therefore likely to vary in their attitude towards process innovations. Managers may continue to purchase machines with which they are most familiar. This would be in line with Leibenstein's contention that monopolistic conditions tend to increase 'X-inefficiency', ie the extent to which costs are higher than necessary.[31] Such X-inefficiency may be a reflection of management's desire for the 'quiet life', financed by the discretionary profits arising from a monopolistic position.

Market imperfections may be felt in other ways. For example, financial resources may be limited and firms may, because of imperfections in capital markets, be unable to raise the necessary capital to install new processes. Furthermore, businessmen may be deficient in their knowledge of the availability of new techniques (see p. 17).

So far we have dealt only with process changes. With product innovations the basic analytical framework outlined by Salter remains the same: under perfectly competitive conditions existing specialist plant will be kept in operation as long as operating costs are covered. New plant will be installed only where expected revenue is sufficient to cover both operating and capital costs. The problem of uncertainty however is likely to be much greater with product innovations.

7.3.3 Some empirical studies of diffusion

A number of studies have been made of the diffusion of both process and product innovations. Griliches for example has provided data on the diffusion of hybrid corn and particular types of agricultural machinery in the United States.[32] Mansfield has analysed diffusion rates for twelve innovations in four American industries.[33] Metcalfe has studied the diffusion process in Lancashire textiles.[34] Ray has reported on an international study of the diffusion of ten process innovations.[35]

For most of these innovations an S-curve describes the acceptance pattern although there are wide variations in the exact nature

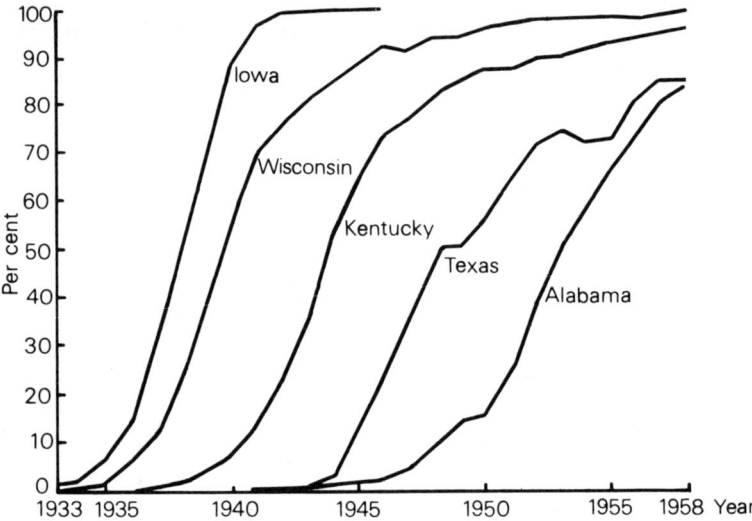

FIGURE 7.2 Percentage of all corn acreage planted to hybrid seed in some US States.
Source: Z. Griliches 'Hybrid Corn: An Exploration in the Economics of Technological Change' *Econometrica* (1957).

of this curve.* Differences occur between innovations, and between areas, for the same innovation. Differences may exist in the dates at which innovations are first employed, the rate at which they are diffused once accepted and the final equilibrium level of penetration (in output terms) that they achieve. Differences in these respects between areas of the United States for one basic innovation are brought out very clearly in the study by Griliches, as shown in figure 7.2. Variations in diffusion patterns between *countries* is shown in table 7.4, which is taken from Ray. It is clear from both the diagram and the table that, apart from a few exceptions, innovations usually take a considerable time to achieve substantial penetration. It is perhaps also worth noting here that, although the UK was the first to adopt six of the ten for which data are given in table 7.4, its rate of diffusion was sometimes among the slowest.

The underlying rational for the S-curve lies in the learning process that industries as a whole undergo in relation to an innovation. Initially little is known about an innovation's capabilities and it is difficult to get additional firms to utilise it. However, as the number of users increases, so the information available increases and the rate of diffusion rises (or, in the case of an unsuccessful innovation, slows down). At the same time (as Salter's model suggests) *competitive* pressures to install the innovation increase. Eventually only firms that (for various reasons) are most opposed to change will remain and the acceptance rate will slow down.

Why do the S-curves vary in shape between innovations and between different regions (and countries) for the same innovation? Griliches's study throws some light on this question. Firstly, in

---

* The studies by Griliches, Mansfield and Metcalfe suggest that a logistic curve fits the data on diffusion fairly well. The equation for this curve is:

$$Y = \frac{E}{1 + e^{-(a + bt)}}$$

Where Y is the percentage of an industry's output produced using the innovation, E is the eventual equilibrium level of diffusion, and t is the time variable. The curve is asymptotic to zero and K, and is assymetric around the inflection point. Griliches takes the linear transform of the above:

$$\log_e \frac{(Y)}{(E - Y)} = a + bt$$

to estimate the parameters directly by least squares. Ray on the other hand found that his data did not provide 'evidence convincing enough for unconditionally accepting [the logistic curve]'.

Table 7.4
Diffusion of ten techniques[a]

	OXY	CC	SP	NC	SL	FG	TK	SCM	ATL	GA
Year of introduction (19...)										
Austria	52	52	66	63	61		57			
France	56	60	65	57	53/54	66	49	60	47	66
Germany	57	54	65	62	54	66	59	53	54	
Italy	64	58	65	60	60	65	51	62	50	
Sweden	56	63	63	58	57		48	50	55	59
UK	60	60	64	55	58	58	02	50	47	59
Number of years after pioneer										
Austria	0	0	3	8	7		9			
France	4	8	2	2	0	8	1	10	0	7
Germany	5	2	2	7	1	8	11	3	7	
Italy	12	6	2	5	6	7	3	12	3	
Sweden	4	11	0	3	3		0	0	8	0
UK	8	8	1	0	4	0	...[b]	0	0	0
Years to produce indicated percentage of output by process (%)	20	1	10		2		10		30	50
Austria	2	10	1	..			4			
France	12	..	2	..			12		..	..
Germany	8	9	2		6		2		..	..
Italy	2	7	..		3		10		15	
Sweden	9	3	2		9		8		2	3
UK	5	6	3		6		..[b]		10	4
Diffusion (%) by indicated year[c] (19...)	67	66	68	66	68	66	66	66	68	68
Austria	67	1.2	35	..	(5.0)	—	58	—	—	—
France	17	0.6	(25)	0.81	(8.5)	7	31	68	..	..
Germany	32	2.4	15	0.35	9.5	6	48	66	81	—
Italy	27	2.0	4	0.36	3.0	6	45	48	38	—
Sweden	33	2.2	52	..	2.4	—	59	80	97	48
UK	28	1.6	24	0.88	8.0	25	12	36	52	70

(a) Processes are abbreviated as follows: OXY: basic oxygen process in steelmaking; CC: continuous casting of steel; SP: special paper presses; NC: numerically controlled machine tools; SL: shuttleless looms in cotton weaving; FG: float glass; TK: tunnel kilns in brickmaking; SCM: new steel-cutting methods in shipbuilding; ATL: automatic transfer lines for car engines; GA: Gibberellic Acid in brewing/malting. FG has not been used in Austria or Sweden, nor GA in Austria, Germany or Italy. There is no industry in Austria comparable with those in the other countries which use SCM or ATL.

(b) Omitted as extreme.

(c) Except for machine tools (number per thousand tools, including aircraft industry), proportions are based on respondents' total output for SL, TK, SCM and ATL; otherwise on national output. Figures in brackets are estimated.

Source: G. F. Ray 'The Diffusion of Technology' *National Institute Economic Review* (1969)

relation to the date at which the diffusion of particular innovations begins, he has shown that hybrid corn entered those areas that had the highest profit expectation first. (As we have seen (fn. on p. 183) the particular S-curve that Griliches uses – the logistic curve – is in fact asymptotic to zero. He therefore takes the date at which a region first planted 10 per cent of the total acreage that was eventually planted with hybrid seed as the date of origin.) It should perhaps be noted that hybrid corn was 'the invention of a new method of innovating, a method of developing superior strains of corn for specific localities'. The actual breeding of particular hybrids had to be done separately in each area. Thus the beginning of the diffusion process in each region can be interpreted in terms of when the right hybrid seed became *available* in that region. Profit expectations of seed producers depend on the eventual size of the market, innovating (including marketing) costs and the expected rate of diffusion. Secondly, he found that the rate of diffusion, *once hybrid corn had been introduced into an area* was in part also determined by the expected profitability to farmers of making the change to the new seed. The supply of corn seed was not a limiting factor, the main determinants of the diffusion rate being on the demand side. Griliches argues that the higher the profit expectation, the faster is adjustment to the innovation undertaken, not only because of the incentive effects but also because information about an innovation is likely to spread more rapidly among potential users the higher the likely returns of this innovation are.

Thirdly, he found that different areas settled at different equilibrium levels of 'penetration' by hybrid corn again because of differences in average profitability of the shift to hybrid corn. Where corn production is in any case a marginal activity, the technical superiority of hybrid corn is unlikely to make as great a total penetration as in an area where such production is already highly profitable. Griliches's emphasis on profitability as a determinant of diffusion rates is supported by the study by Mansfield.[36] The latter has also shown that, *for equally profitable innovations,* the probability of adoption is likely to vary inversely with the size of the investment required because firms will be more cautious in committing financial resources to big projects. They may also experience greater difficulty in raising the necessary funds.

7.3.4 *Leaders and followers*

Although the above work provides a useful guide to what are likely to be important factors in determining the pace of diffusion of an innovation in an industry as a whole, individual firms are likely to vary in their responses to an innovation for a wide range of reasons which may not be directly connected with the factors outlined above. Is it possible to identify the characteristics of firms that are technical leaders? Carter and Williams have provided a list of twenty-four characteristics that are 'common to all or most' of what they have called the technically progressive firm – the firm 'which is in the forefront of discovery in applied science and technology, and which is quick to master new ideas and to perceive the relevance of work in neighbouring fields'.[37] These characteristics are generally absent in other firms. They relate to many different aspects of the firm's operations, including its policies towards capital investment, R & D wherever it is carried on, training (including management development) and other firms. Carter and Williams found that the technically progressive firm was of 'good general quality'. Some of the characteristics mentioned by Carter and Williams may be systematically related to the size of firm. (We are not here examining the relationship between firm size and the *first* firm to use a technique, ie the innovator (see chapter 3) : we are more concerned with the characteristics of firms that tend to be among the earliest adopters, if indeed there is such a 'tendency.') For example, large firms may tend to have more sophisticated appraisal techniques than smaller firms. In his study of selected innovations, Mansfield has in fact shown that larger firms do tend to be quicker in adopting an innovation than smaller ones. Metcalfe's study also emphasised the importance of the large firm in the diffusion process. However this may result not from greater technical progressiveness, but merely from its larger size *per se,* which may make certain innovations more profitable to them than to smaller firms. Ray on the other hand has pointed out that on the basis of the study conducted at the National Institute there is

> ... no definitive evidence for saying that large companies have always been in the forefront of technical progress in the sense of being leaders in innovation and the adoption of new techniques. The leading role which they often play in research and development, their generally more sophisticated managerial set-up and

their easier access to new capital are likely to give them a lead over smaller firms : some of the case studies do indeed point to the outstanding part played by large companies. But in other cases it has been the opposite way round.[38]

Whatever the characteristics of the technically progressive firm, it does not seem as if firms necessarily *remain* progressive for long periods. Mansfield has shown from his studies that 'Even if one firm was considerably quicker than another to begin using one innovation, the chance that it will also be quicker to introduce another innovation occurring only five years later is not much better than 50 :50.[39] In other words, firms do not consistently remain at the forefront of technology'. It is possible however that *for economic reasons* a firm that is first to introduce one particular innovation will not be the first to introduce the next innovation, because the attractiveness of the latter will be less, given that it has already installed more modern equipment.

So far, we have assumed that *existing* firms will be the vehicles for increasing the percentage of an industry's output produced by a new process. However the importance of new firms in the overall diffusion process must not be underestimated. Tilton for example has shown that newly established firms, often founded by people with experience in the industry, have played a major part in the diffusion of semiconductors in the United States.[40]

7.4 Conclusion

Most industrial R & D, whether financed by government or from private sources, is heavily concentrated in a few industries. There are wide differences in research intensities between industries, which can be explained in part at least by differences in the profitability of R & D. This is supported by studies of R & D expenditure at the level of the individual firm. Industries also vary in their commitment to basic research : again, there are sound *a priori* reasons why this variation should exist.

When we turn to patent data a heavy concentration in a few industries is again apparent. The patent-to-R & D expenditure ratio varies widely between industries. This is as expected, bearing in mind that the patentability of R & D in different fields is likely to vary, as is the attitude of industries to obtaining patent protection. However, the patent and R & D data are collected on different bases; hence any analysis must be tempered with caution. As far as royalty and know-how payments are concerned, British industry as a whole is in substantial deficit. However it is difficult from these figures alone to comment on how technologically progressive British, as opposed to foreign rivals are.

Most work done on the diffusion of innovation to date suggests that the acceptance pattern tends to follow an S-curve. A few studies have also shown that the profitability of adopting an innovation is a major determinant of the diffusion pattern of different innovations and of the same innovation in different areas. While it may be possible to identify those firms that are likely to be leaders in the diffusion process, there seems to be no evidence that firms will remain technically progressive for long periods of time; what little evidence we do have suggests that such leadership tends to be short-lived.

NOTES on Chapter 7

1. M. Peck 'Science and Technology' in R. E. Caves (ed.) *Britain's Economic Prospects* Washington, Brookings Institution (1967) p. 482. See also C. Freeman 'Research and Development: a comparison between British and American Industry' *National Institute Economic Review* (1962)
2. J. E. Tilton 'Research and Development and Industrial Growth' *Journal of Political Economy* (1973). See also Leonard's reply in the same journal.
3. *FBI Survey of Industrial Research in Manufacturing Industry* London, FBI (1960)
4. C. Freeman op. cit.
5. ibid.
6. W. N. Leonard 'R & D in Industrial Growth' *Journal of Political Economy* (1971)
7. Tilton op. cit.
8. J. R. Minasian 'The Economics of Research and Development' in National Bureau of Economic Research *The Rate and Direction of Inventive Activity* Princeton, Princeton UP (1965)
9. C. F. Carter and B. R. Williams *Industry and Technical Progress* Oxford, Oxford UP (1959) p. 51
10. H. G. Grabowski 'The Determinants of Industrial Research and Development: a study of the chemical, drug and petroleum industries' *Journal of Political Economy* (1968), reprinted in B. Yamey (ed.) *The Economics of Industrial Structure* Harmondsworth, Penguin (1973)
11. E. Mansfield *Industrial Research and Technological Innovation* London,

TECHNOLOGICAL CHANGE IN INDUSTRY 189

      Longmans (1968) chapter 2
12. ibid. chapter 3
13. Minasian op. cit.
14. Quoted in D. Hamberg 'Invention in the Industrial Research Laboratory' *Journal of Political Economy* (1963)
15. *Industrial Research Survey* London, FBI (1961)
16. J. Rose 'Why Research Pays Off – and How' *ICI Magazine* (1971)
17. *Basic Research, Applied Research and Development in Industry, 1965* Washington, National Science Foundation (1967) p. 78
18. National Science Foundation op. cit. pp. 79–80
19. C. T. Taylor and Z. A. Silberston *The Economic Impact of Patents* Cambridge, Cambridge UP (1973) p. 64
20. *ibid* pp. 112–13
21. J. L. Enos 'Invention and Innovation in the Petroleum Refining Industry' in National Bureau of Economic Research op. cit.
22. J. M. Blair *Economic Concentration* New York, Harcourt Brace (1972) p. 228
23. *ibid.*
24. US National Commission on Technology, Automation and Economic Progress, Technology and the American Economy, Vol. 1 (1966) pp. 3–4. Quoted in D. S. Landes *The Unbound Prometheus* Cambridge, Cambridge UP (1969) p. 519
25. See J. Downie *The Competitive Process* London, Duckworth (1958) and W. E. G. Salter *Productivity and Technical Change* 2nd ed, Cambridge, Cambridge UP (1966)
26. W. E. G. Salter op. cit.
27. ibid. p. 52
28. R. G. Gregory and D. W. James 'Do New Factories Embody Best Practice Technology?' *Economic Journal* (1973)
29. ibid.
30. G. F. Ray 'The Diffusion of New Technology' *National Institute Economic Review* (1969). While the final draft of this book was being prepared, the full report of the study on which Ray's article was based was published: see L. Nabseth and G. F. Ray *The Diffusion of New Industrial Processes* Cambridge, Cambridge UP (1974)
31. H. Leibenstein 'Allocative efficiency v. X-efficiency' *American Economic Review* (1966)
32. Z. Griliches, 'Hybrid Corn: An Exploration in the Economics of Technological Change' *Econometrica* (1957)
33. E. Mansfield *Industrial Research and Technical Innovation*, London, Longmans (1968) Part IV
34. E. Metcalfe 'Diffusion of Innovation in the Lancashire Textile Industry' *Manchester School* (1970)
35. G. F. Ray op. cit. See also fn. 27
36. E. Mansfield op. cit.
37. C. F. Carter and B. R. Williams op. cit. p. 178
38. G. F. Ray op. cit.
39. E. Mansfield op. cit. p. 172
40. J. E. Tilton *The International Diffusion of Technology: The Case of Semi Conductors* New York, Brookings Institution (1971)

8 Co-operative Industrial Research*

For over fifty years a government scheme to give financial aid to industries organising research and technical activities on a co-operative basis has been in operation in the UK.[1] The scheme was established in 1917 by the (then) newly established Department of Scientific and Industrial Research (DSIR) to meet the acute technological needs of British industry revealed by the First World War. It was originally intended that the grants, which were based on the grant-earning subscriptions of UK members to the programme, should be discontinued once a research association (RA) had found its feet. The interwar period revealed however that a policy of disengagement would have resulted in the collapse of most of the RAs. After 1945 this policy was formally abandoned, and the government grant became a more permanent feature of the scheme. A number of European countries now have similar schemes, modelled in part on the British pattern as shown by table 8.1.

Unfortunately, 1962 is the latest year for which comparable data are available. There is no government-supported system of co-operative industrial research in the US, although there are a number of co-operative bodies whose aim is to support research on a collective basis, either in their own laboratories or, as frequently, by sponsoring a research programme in a contract research institute or university.

---

* This chapter is based on a fuller study by the author into the operation of co–operative research associations in British industry. The full findings of the study are found in his 'The Role and Operation of Co-operative Research Associations in British Industry', Ph.D. thesis submitted to Nottingham University, 1970. A shortened version of this has appeared as *Co-operative Research in Industry: An Economic Study* London, Martin Robertson (1973).

Table 8.1
RAs in Europe, 1962

	No. of laboratories of each type			Sources of income (% of total)		
	Grant-aided	Non-grant-aided	Total income, 1962 (£m. p.a.)	Govt	Industry	Contract research
Austria	22	—	0.425 (1961)	10	50	40
Belgium	55	—	2.16 (1961)	31	57	12
Denmark	5	6	0.47	9	77	14
France	—	89	16.7 (1960)	2	90	—
Germany	61	—	4.58	30	63	7
Italy	4	—	0.082	50	50	—
		8	n.a.	—	100	—
Netherlands	18	—	2.80	60	7	33
Norway	3	15	0.50	10	90	—
Spain	9	—	0.112	50	50	—
Sweden	5	12	1.53	20	75	5

Source: *The Structure of Industrial Research Associations* Paris, OECD (1965)

It is the object of this chapter to examine the current role of the RAs in British industry. In the first section the scale of the RAs' activities in the UK is discussed. The second section examines the economic basis of co-operative research. The activities of the RAs are then analysed in the light of this discussion in the third part of the chapter. The last section deals briefly with the financing of the RAs.

8.1 The present scale of RAs in the UK

8.1.1 *The RAs in the overall R & D effort*

At the end of 1971 there were 39 government grant-aided RAs, covering most sectors of British industry and having over 20,000 members, 93 per cent of which were UK Ordinary or Associate Members. (Some firms of course belong to more than one RA.) A few co-operative research bodies, like the Permanent Magnet Association and the Tobacco Research Council, operate outside the government scheme.

The RAs' share of the UK's total R & D effort has been very small, and may even have declined slightly over the period 1955/56

– 1970/71. They accounted for only 1.5 per cent of the total R & D carried out in 1969–70, compared with industry's 59 per cent.

8.1.2 *The RAs in individual industries*

When the importance of an RA in its own industry's R & D expenditure is considered, however, the picture presented is in many cases very different. It can be seen from table 8.2 that in certain sectors of manufacturing, eg shipbuilding and marine engineering, clothing, footwear and leather, the RAs account for a significant percentage of their industry's total R & D expenditure. At the other extreme, the Electrical RA is responsible for less than one per cent of its industry's total R & D.

Table 8.2
Expenditure on co-operative research in British manufacturing industry, 1969–70

	(1) RA expenditure (£m.)	(2) Total R & D expenditure (£m.)	(3) (1) as % of (2)
Food, drink, tobacco	8.1	18.3	4.4
Chemicals	1.0	86.8	1.2
Iron and steel	0.8	9.2	8.7
Non-ferrous metals	0.7	6.9	10.6
Mechanical engineering	3.0	55.3	5.4
Scientific instruments	0.8	16.1	5.0
Electrical engineering	1.1	171.5	0.6
Ships and marine engines	1.0	4.8	20.1
Motor vehicles	0.4	40.3	1.1
Aerospace	—	158.9	—
Railway equipment	1.0	2.4	—
Textiles, man-made fibres	1.1	13.9	7.8
Leather, leather goods, fur	0.1	0.3	34.3
Clothing and footwear	0.3	1.0	29.1
Building materials, etc	0.5	4.0	11.7
Pottery, china, clay	0.3	6.1	5.3
Timber and furniture	0.5	0.9	52.2
Paper, printing, etc	0.4	4.8	9.2
Rubber and rubber products	0.5	4.3	10.9
Other manufactures	0.2	4.3	0.4
All manufactures	13.8	620.3	2.2

Source: *Research and Development Expenditures* (Central Statistical Office) London, HMSO (1973) table 17D

The importance of an RA or RAs in an 'industry' will depend firstly on the definition of the industry to which the RAs are initially allocated, and secondly on the way the industries with which they are compared are themselves grouped. The two problems are closely interlinked. The first presents few difficulties that are not encountered in any industrial classification. No RA's work is limited to one isolated sector; eg the Production Engineering RA is classified here under Mechanical Engineering, although its work will be of relevance to other industrial sectors. Similarly, the Electrical RA is classified under Electrical Engineering, but its work has obvious implications for other industries, such as the motor and shipbuilding industries. Research in the Welding RA (now the Welding Institute) involves work on the properties of metals, both ferrous and non-ferrous, but as welding is essentially a process it has been classified under Mechanical Engineering.

The second problem involves more complex issues; for example, in the chemical industry certain RAs, eg the Paint RA, serve specific sections of the industry. Should the income of these RAs be set against the R & D expenditure of the chemical industry as a whole, or of only those sections directly concerned? Or again, should the Motor RA's income be compared with the R & D expenditure of the whole vehicle industry including railway equipment and possibly aircraft, or just the motor vehicle sector as defined under the Standard Industrial Classification? In both cases, different percentages would result from taking different industrial boundaries.

One area in which the classification problem is acute is the textile industry, where man-made fibres now form a significant part of the market. This latter sector of the industry is far more 'science-based' than the hosiery, cotton and wool sectors, which still depend very heavily on traditional crafts and skills. All the RAs in the textile industry, however, are devoting an increasing proportion of their resources to examining the implications of the use of man-made fibres for their own trades. Figures published by the FBI in 1960 (there are no similarly detailed figures available for later years), showed that the RA's expenditure as a percentage of all R & D was high in all sectors except man-made fibres:[2] 55 per cent in cotton; 59 per cent in hosiery; and 84 per cent in wool. When these figures are placed together with the man-made fibres sector, where only 7 per cent of its total expenditure was channelled through the RAs in 1960, the weighting of the latter sector is very heavy, resulting in an overall figure for textiles of only 16.6 per cent in 1960 and 7.8 per cent in 1968–69. This problem is also marked in paper and printing. In the paper and board industry only 6 per cent of its total R & D expenditure in 1960 was made by its RA. (The Paper and Board RA has now amalgamated with the Printing and Allied Trades RA.) But in the printing industry the RA spent 50 per cent. When the two are brought together, the overall figure for 1960 was only 11.7 per cent.

It should be noted that, even given the statistical difficulties discussed above, the figures in table 8.2 over-estimate the extent of the industry's own support of co-operative research for two main reasons. Firstly, as noted earlier, a proportion of several RAs' expenditure is on contract work, where the results may be confidential to the sponsor, or on other non-research activities (see below). Hence the percentages do not reflect accurately the importance of co-operative work in an industry's R & D. Contract research is particularly important in the RAs in the electrical, mechanical engineering and scientific instrument industries. Secondly, with nearly all the industries shown the RA income and the industrial expenditure on R & D include a proportion of government money. Hence one cannot say that the percentages calculated in table 8.2 indicate the extent of an industry's own *independent* evaluation of the RA's work. In all except a few cases it is over-estimated by the amount of government grant, and probably more, as the latter acts as an inducement to additional subscriptions over and above those that would have been obtained in the absence of the grant.

Table 8.2 does not cover all the RAs in existence, as the figures concern only manufacturing. Grant-aided RAs also serve the coal and coke, laundry, construction and, until recently, the water supply industries.

Why do variations in the importance of RAs in manufacturing occur? One explanation may be that in the more science-based industries, where firms are heavily dependent on scientific and technological advance for their competitive strength, the co-operative RA will be of less relative importance than in those industries that are still very largely traditionally based. Most firms in the former have their own R & D laboratories and it will be to these laboratories that they will primarily look for new and improved processes and products. In such industries the RA is likely to play a fringe role, acting as a 'listening post' and performing 'background' research. In the traditional industries however the RA may often be ahead of its members technologically and is likely to be of much greater importance in the industry's overall R & D commitment. This finds support from the OECD's view that such industries 'as have a large number of component members and whose overall research effort is not highly developed are likely to be most suitable for RAs'.[3]

There does in fact appear to be a relationship between an industry's science base, measured by the number of QSEs engaged on R & D per one hundred of all employees, and its proportionate support of its RA, measured by RA income expressed as a percentage of the industry's total R & D expenditure: the correlation coefficient, using the FBI's 1960 data, is -0.59, significant at the one per cent level. There are clearly other factors at work, eg the size structure of the industry concerned (this may not however be independent of the industry's science base), and the data for reasons outlined above are very crude.

In some industries that are traditionally based and carry out very little R & D, management attitudes may still be very much opposed to co-operative research. For example, Carter and Williams found that in the cutlery industry, 'The structure of the industry, the strong handicraft tradition, the fierce competition when orders are short, have not encouraged a co-operative spirit . . . Individual cutlery manufacturers have tended to watch competitors with suspicion and to guard jealously any novel ideas which they thought might assist them to get the better of their competitors.'[4] These attitudes may go a long way to explain the fact that, although there were only 0.22 QSEs working on R & D per one hundred employees in 1960 in 'Other Metal Goods', only one per cent of that sector's R & D was carried out in the relevant RAs. In some industries, management may be so traditional that it opposes not only co-operation in research, but research itself.[5] In the absence of these circumstances, however, it is reasonable to suppose that the more science-based an industry is, the less important is an RA likely to be.

8.1.3 Membership of the RAs

As we have seen, some RAs, eg the Glass Industry RA, serve identifiable industries while others cater for several. Perhaps the best example of the latter type is the Welding RA, which serves a wide range of industries. Welding equipment manufacturers who might have been thought to have the closest interest in the RA's work provided less than ten per cent of all co-operative income in 1965. Even in the glass industry closer inspection reveals a diversity of interests : Pilkington's (the flat glass manufacturers), the container manufacturers and the crystal glass firms, who together constitute over eighty per cent of the RA's industrial income, have very little in common technically beyond the glass melting stage. Such diversity of interest inevitably causes problems in the drawing up of a co-operative research programme. It may lead to the RA's providing a broadly based programme which in turn leads to the spreading of resources too thinly over too many projects. Some RAs, eg the Launderers' RA, have both the manufacturers and users of goods and/or equipment in membership. Conflict between these interests can also occur. Divergence of interest may also exist between large and small members. The small member firm is likely to be more interested in shorter-term work of a trouble-shooting nature, whereas the larger firm may be more anxious for the RA to do more longer-term, basic research, which complements its own R & D.

This difference in need came out clearly in the author's interviews with industrialists on the role of the RAs. For example, 'When you are in trouble you go to the Doctor', was how one director of a small company likened his relationship with the Glass Industry RA. 'You know they are always there' was the justification for membership of the same RA in another small company. A technical director of a large company, however, summed up its relationship with the RA as follows : 'The RA's work rarely has any immediate and direct effect, but we develop the ideas that come out of the RA into the production processes.'

8.1.4 *The Size of the RAs*

The RAs vary considerably in size, as is shown in table 8.3. Half the RAs were operating on budgets under £300,000 in 1970; four had incomes of under £50,000. (By 1973, only one of these, the Cutlery RA, remained as a separate grant-aided RA.) Because costs per research worker differ substantially from industry to industry it is not possible to say in general terms what is too small. This was

Table 8.3
The size of RAs, 1970

Size (income) (£000s)	No. of RAs	Total income
Under 50	4	136,967
51–100	3	250,635
101–150	4	522,865
151–200	5	865,306
201–300	5	1,280,511
301–400	7	2,340,695
401–500	8	4,088,936
601–800	2	1,374,503
801–1000	1	853,625
over 1000	4	5,418,322
	43	17,132,365

Source: Department of (Trade and) Industry

shown clearly in the mid-1960s, when both the Hat and Allied Feltmaker's RA and the Internal Combustion Engine RA left the government scheme within a year of each other on incomes of £40,000 and £108,000 respectively. Amalgamation of RAs working in similar fields has been encouraged in recent years and this process might justifiably go further to obtain economies in the use of indivisible equipment and of common services. It should be remembered, however, that too little is known about the relationship (if any) between creativity and organisational size for a strong case to be made for bigger units on these grounds alone. Furthermore, many of the comparable units in other European countries, especially in Norway, are small compared with British standards.

8.2 The economic basis for co-operative research

8.2.1 *The R & D spectrum*

As shown in chapter 1, the results of basic research are usually freely and widely available. The social returns from such work are likely to diverge widely from the private returns accruing to the individual firm undertaking it, since basic research results cannot be patented. On these grounds, co-operative rather than individual effort in this field is likely to be encouraged. But the uncertainty of the outcome of this type of research and the normally long pay-off period required before it becomes profitable, if at all, still remain as disincentives to co-operation in basic research. The fact however that firms will be risking less by sharing the cost of a project may tend to reduce this reluctance. The net effect of these forces on the willingness of a firm to support co-operative basic research cannot be deduced from a *priori* reasoning, especially as it may be determined in part by the number of firms involved. *How* basic the project is, the general attitude of industry to research, and the closeness of product and process technology to basic research findings will also affect this willingness.

Firms as we have seen normally utilise the majority of their resources in applied research and development, since it is these spheres rather than in basic research that a firm's survival and profitability will be at stake. The economic characteristics of applied research and development also suggest that they are generally unsuitable activities for co-operative work. However co-operation will be attractive to individual firms even in this area whenever the advantages of cost reduction made possible by such co-operation more than outweigh the adverse effects on demand which result from the total market being shared out between the participating firms.

Competition has been examined here in an inter-firm context. But if an inter-*industry* context is taken, firms may be prepared to co-operate nearer the market end of the spectrum even in a competitive industry in order to present a united front to other industries in the same country or, taking a wider context, to foreign companies seeking outlets in the home market. In both these cases however, advantages of collaboration in other spheres, eg marketing, may be strong enough to induce complete integration rather than just co-operation in research between firms.

If economies of scale exist in R & D and particularly in development, then co-operation between firms may be necessary, although once again outright amalgamation may be a more satisfactory solution because of economies elsewhere. (This possibility is reinforced by the fact (p. 64) that R & D costs are often only a small percentage of total innovation costs.)

8.2.2 *Restrictive practices and co-operative research*

In the light of the above discussion, it is worth noting that a number of trade associations have claimed, in support of their price agreements, before either the Monopolies Commission or Restrictive Trade Practices Court that collaboration in R & D or the exchange of technical information is fostered by the more secure atmosphere engendered by a price agreement (see p. 84). In the absence of such agreements, however, it seems from the nature of the R & D spectrum that there will be a 'pull' away from co-operation in basic research on the one hand, and from applied research and development (particularly the latter) on the other.

There would therefore be a point beyond which the individual firm would not be prepared to go in supporting a co-operative programme. This limit would be affected *inter alia* by the intensity of competition and the level of indivisibility in R & D. The strength of the pull at both ends of the spectrum is continuous in nature; hence the pull away from co-operation at either end will increase in intensity as a project becomes more basic or more market-orientated. It is instructive to note that in Britain and France, two countries for which figures are available, the majority of the RAs' resources is spent on applied research – in the middle of the R & D spectrum. And as shown below, many RAs in the UK devote a considerable percentage of their resources to non-competitive topics.

8.3 The RAs' activities

While the RAs as a whole have several important inventions to their credit – the Chorleywood Bread process from the Baking Industries' RA and BISRA's* Fuel–Oxygen–Scrap process are only two examples – they have normally performed a much less spectacular role. In the OECD's view, 'the RAs' greatest service has been perhaps the change they have accomplished almost imperceptibly in the scientific outlook of the industries they serve'.[6] This has been particularly true of some of the RAs serving traditional industries.

*British Iron and Steel RA, now the group laboratories of the British Steel Corporation.

In the laundry industry for example, the British Launderer's RA has over the years scientifically investigated long accepted 'rule of thumb' measures for the use of water heat and detergents in the washing processes, and has come up with recommendations leading to considerable savings in power and materials. In 1954 the RA with some justification claimed that 'practically all washing processes used in this country today are based on BLRA recommendations'.[7] This in turn has made the average launderer much less sceptical of 'boffins in white coats'. Other RAs, eg the Welding RA, have acted in a 'midwifery' role by testing and evaluating new methods, sometimes improving them, and then introducing them to the industry. Most of the RAs have been heavily engaged in testing and evaluating new materials and processes and in providing new testing methods. They have also undertaken basic research studies. For example, the Food RA has conducted investigations into the physical properties of certain fat mixtures, and in the printing industry the RA has examined the fundamental properties of ink and its behaviour on rollers.

Although the picture inevitably varies from RA to RA, by and large the RAs have not been *engaged* in research that could have led to important and immediate commercial repercussions. One RA director has suggested a reason for this: 'there is a tendency for RAs to be left with longstanding and refractory problems which inidividual firms have failed to solve; or else the problems put by industry are those whose solutions, while generally desirable, have no spectacular impact'.[8] It may be argued however that it is not an RA's function to undertake research of direct commercial relevance to individual members since this would involve treading on competitively sensitive toes; rather that their objective should be to concentrate on 'complementary' areas of work where members may be happy to collaborate, such as measurement and testing research, research directed towards improving industrial health and safety, the evaluation of new and existing processes, education, information and advisory services, and the provision of specialist services too costly for any one firm to maintain.[9]

In table 8.4, the expenditure patterns of the thirty five RAs for which figures were available for 1970–71 are given. It can be seen that, for the 30 RAs for which separate figures are provided, 16 spent less than 50 per cent on co-operative research (excluding co-operative research on standards). Most of those who spent over

50 per cent operated in 'traditional' industries. It would have been useful if a further breakdown of the co-operative research programmes had been available, to assess how far the RAs' co-operative research was aimed at producing new products and processes for commercial use, and how far it was aimed at non competitive topics, such as health research, and at basic research. Unfortunately this information is not available. The table also shows the commitment of most RAs to information and liaison activities, activities that are very difficult to evaluate in economic terms. Most RAs spent at least 10 per cent of their expenditure in these fields and seven spent over 20 per cent.

As suggested in section 8.2, basic research may provide another area in which firms may be more willing to co-operate, although the distance and uncertainty of the pay off must be offset against the lower research cost for each individual member. Furthermore the RAs compete here with the universities and (to a lesser extent) government laboratories. Since 1969 the government has attempted to steer the RAs away from basic research to more applied research and to areas such as 'industrial and management training, operation research, ergonomics, inter-firm comparison, production economics and market research of a technico-economic nature'.[10] This is probably a correct policy in view of the RAs' industrial accountability.

Thus, although there may be areas of complementary activity well suited to a co-operative research body, the fact that this work has not usually led to commercially important inventions may affect the attitude of industry to the RAs. Companies may feel that an RA's work requires only occasional surveillance and not continuous involvement. Since the advantages of RA membership cannot be clearly seen by a list of their inventions, a superficial view may tend to discount their value. Many industrialists have stressed to the author the impossibility of quantifying the benefits of RA membership.

8.3.1 *A survey of the views of industrial research directors*

In 1969 a questionnaire was sent out by the author to research directors in industry asking them for their views on the role of the RAs. The detailed analysis of the findings of this survey and its limitations, which are readily acknowledged, are found elsewhere,[11]

Table 8.4
Breakdown of estimated expenditure of RAs (by percentage of total expenditures 1970–71)

RA	Co-operative R&D (excluding standards) (%)	Co-operative R&D for standards (%)	Non-R&D work on standards (%)	Sponsored R&D (%)	Non-R&D technical work (co-operative or individual) (%)	Non-R&D management work (co-operative or individual) (%)	Publications and information (%)	Routine liaison visits (%)	Training and education (%)	Other (%)
Brush	52.1	—	—	3.2	8.5	—	18.4	10.8	7.0	—
Cast Iron	33.9	—	—	—	—	16.5	12.2	—	0.9	0.3
Ceramics	55.4	5.2	2.6	12.9	36.1	—	4.5	4.4	0.9	—
Coke	85.0	—	—	—	14.2	—	1.1	0.2	—	—
Cotton	50.6	0.9	—	25.8	13.7	0.7	8.1	—	2.8	2.9
Cutlery	68.1	—	—	—	8.4	—	12.7	—	—	—
Drop Forging	68.5	—	3.4	—	19.2	3.4	9.5	2.7	0.3	—
Electrical	32.4	0.7	0.9	48.0	13.2	5.6	0.3	—	—	0.7
Fruit and Vegetable	——— 78.0 ———				15.0	—	5.0	—	2.0	—
Furniture	24.3	—	1.5	3.1	14.1	21.4	11.6	1.2	13.7	8.9
Glass	64.4	1.5	0.2	5.7	9.2	4.2	11.2	1.7	1.0	0.9
Hosiery	57.2	0.9	0.6	1.9	13.1	—	7.0	6.8	12.3	—
Hydromechanics	44.3	—	—	23.9	—	—	28.2	2.4	1.2	—
Industrial Biology*	——— 80.0 ———				—	—	19.5	—	—	0.5

Industry										
Jute	81.2									7.6
Linen*	66.5									
Lace*	76.1									
Launderers'	38.6	—	—	—	—	—	11.2	—	—	—
Leather	55.7	1.2	0.6	—	25.7	10.4	8.9	—	5.7	0.8
Machine Tools	49.0	1.0	1.2	5.2	15.8	1.1	23.9	9.4	1.4	10.3
Motor	76.0	10.3	—	12.0	14.8	1.2	10.8	3.5	0.2	—
Non-Ferrous Metals	34.4	—	0.1	2.4	1.3	—	4.6	2.6	—	—
Paint	35.0	0.9	1.1	37.0	16.0	2.1	19.1	2.9	0.1	—
Paper, Board Printing and Publishing	38.2	1.7	1.7	26.3	17.3	—	9.9	1.3	2.3	—
				4.5	23.4	3.1	7.2	3.1	8.5	
Production Engineering	10.4	—	—	—	29.5	13.8	15.8	4.6	16.5	—
Rubber and Plastics	40.6	0.5	2.0	4.4	20.3	1.4	15.8	—	—	—
Scientific Instruments	49.2	0.6	—	31.1	9.5	0.9	25.1	—	—	—
Ship	60.6	4.1	0.5	1.4	16.4	4.4	30.7	1.6	0.6	4.4
Shoe	27.3	—	1.2	1.2	3.0	19.2	8.8	6.5	7.4	5.6
Spring	43.7	0.6	—	20.1	10.3	1.6	6.0	11.9	—	—
Steel Castings	71.6	—	—	—	10.7	—	28.5	3.7	1.8	—
Tar	46.8	7.6	2.0	12.4	15.2	3.8	11.9	0.7	—	2.4
Welding	37.7	0.6	—	26.7	10.3	3.8	12.0	—	10.7	—
Welwyn Hall	84.6	—	—	—	9.2	—	8.7	1.5	—	—
Wool	52.3	1.4	3.4	1.5	9.7	—	7.8	—	8.5	6.1
							4.7			
							17.2			

* 1969 figure

Source: Department of Trade and Industry

but some of the general conclusions are of interest here. (It should be noted that the survey was primarily concerned with research directors in larger firms.)

Firstly, nearly all the directors who replied considered they were obtaining 'value for money' from their RA subscriptions, although they were mostly unable to quantify it. This response would of course be expected from the fact of membership, but it does indicate that at least some industrial research directors see the RA type of organisation as performing a useful role. Secondly, although the areas in which directors considered they were getting value for money varied, it was clear that the majority saw the RAs' *non-*research activities and their research into commercially less important areas as their main value. Less than one third of the directors who replied could name (or wanted to name) 'important cost-reducing benefits' derived from the RA, and again, only a third named any R & D projects that had been the result of 'information, example or idea' passed on by the RA. Only 4 of the 28 directors who indicated activities in which they thought value for money was obtained specifically mentioned the *research* activities of the RAs. Thirdly, most companies wanted the co-operative research activities of the RAs to be maintained.

It is clear then that, despite some of the difficulties the RAs face, there are areas that are considered suitable for co-operative research or technical work. Economic argument over the use of indivisible equipment and services – the Motor Industry RA's proving ground and the provision of computer services for the smaller firm by the Shoe RA are two examples here – would also lead us to expect this. It is interesting to note the concluding comments of the chairman of the FBI Conference on R & D in 1962 : 'The RAs have come out of this conference well; it is clear that they have an important series of roles to play in providing information services, in assisting the small firm and traditional firm in a variety of ways and in basic research.'[12] The omission of applied research and development for the large firm in this statement is noteworthy, especially in view of the findings of the survey outlined above.

8.3.2 *Contract research*

Sponsored research may provide one area of expansion for at least some RAs. Although conflict between co-operative and contract programmes is possible, experience in many RAs has shown the practicality of running both in parallel. Contract research however requires an orientation in attitudes and organisation that has often been lacking in the management of co-operative research. Furthermore, it would not be suitable for all RAs, since the sponsoring of research requires a technical sophistication on the part of firms financing it which is lacking in several RA memberships. Nevertheless, in 1970, eleven RAs were obtaining over 20 per cent of their income from contract research, and that number is probably larger now.[13] It should be noted however that 63 per cent of all contract income went to four RAs in that year. Closer inspection of the figures also indicates some dependence on government contracts: a third of the RAs' contract income came from the government (over half in 1968). This dependence is higher for the smaller RAs. Currently one-half of the RAs' contract income comes from industry, and if the US is any guide, the market for industrial contract research is by no means exhausted. Nearly 75 per cent of this however goes to the four RAs already mentioned. The smaller RAs therefore appear to be somewhat less able or willing to obtain their contract income from industry. It should be remembered that the RAs have to compete here with the independent research institutes.

8.4 Financing of the RAs

8.4.1 *Industrial subscription*

Subscriptions to the co-operative programme are usually paid by member firms on the basis of some size criterion (most frequently, turnover), and in relation to the relevance to them of the programme. In the majority of cases subscriptions are paid into the general pool of funds and are not allocated to particular areas of work, although in an increasing number of RAs members can specify (within limits) the research to which they wish their payments to be directed. In cases where general payments are made, however, members 'lose sight' of their subscriptions on payment.

In view of the diverse interests of RA members outlined in the first section, a better method of industrial financing might be to allow members to allocate most of their subscriptions to particular research areas. Obtaining a greater specificity in industrial subscriptions in this way might focus industry's attention more on the RA's programme and would work in combination with any change in government grants (see below). The encouragement of multi-client projects – a half-way house between general co-operative and individual contract work – and the introduction of charges for liaison services in recent years has gone some way towards meeting this objective, although there is still scope for further action. Members might make a general subscription for maintaining the common services of the RA and some limited basic work, but beyond this members would contribute to specific projects only. Not only might this help to overcome the problems of the diversity of members' interests in many RAs, but it might also increase the sense of commercial accountability of RA staff, and the commitment of companies to the RA's programme, since their financial support would be for identifiable projects.

8.4.2 *Government grant*

The government grant is payable within specified limits on all income that can be classified as grant-earning. A block grant is payable on a specified minimum industrial income, after which an additional grant is given up to a certain maximum, on the basis of any *additional* industrial income above the minimum. Although the terms have been progressively tightened over the postwar period, there is evidence that for some RAs it may have been given too lightly : eleven RAs have consistently earned the maximum over at least a substantial part of that period.[14]

Although through the grant the government has reserved the right to involve itself with an RA's programme and organisation (it has for example insisted that some attempt should be made by RAs to evaluate, in economic terms, its proposed programme of research), it has at the same time emphasised the autonomy of the RAs. The possible conflict between these principles was brought out, perhaps unintentionally, in 1961 by the then Secretary of the

DSIR: 'The autonomy of the industrial research association has been the cornerstone of the whole scheme . . . Nevertheless, the Department by planned selective additional spending, *can transform the whole nature* of the research programmes of the associations'[15] (author's italics).

If it is accepted that the government grant should be made only to support activities that would not otherwise be undertaken by industry (see p. 125), or to speed up research that would otherwise be pursued more slowly, then the grant should be designed to attract the maximum contribution that industry is willing to pay for the research programme. A grant will be justified, therefore, where the anticipated social cost–benefit ratio makes a research programme socially desirable, but where the (perceived) private return to industry may be insufficient to induce the latter to undertake it in part or in full. It is likely however that projects will vary in their private and social cost–benefit ratios. If this is so then some greater specificity in the grants might be an improvement. Grant rates would then vary from project area to project area. For some projects, no grant might be payable. A limited amount of free-ranging research would be financed by a research 'overhead' charged on all projects. A small government grant might also be given. A balance would of course have to be sought between the benefits of applying this method and the administrative costs involved. For a few RAs serving technologically backward industries, a general grant may still be the best form of assistance.

8.5 Conclusion

Although the RAs are only a small part of the country's overall R & D effort, they are often very important in individual industries, especially those that are traditionally based. Membership of the RA often covers a wide variety of technological interests and this may create problems in the determination of the co-operative programme. The RAs vary widely in size, but costs per research worker also vary so considerably between industries that attempts to make a generalised case for a minimum viable size must be treated with caution. Economic analysis would suggest that the co-operative form of research organisation operating in a competitive environment is likely to operate best in applied research and in 'complementary' areas of research which are not commercially sensitive. Individual RAs have made important specific technological innovations, but by and large their work has been of an unspectacular, 'little by little' nature. Nevertheless, there can be little doubt that the co-operative RA is, within limits, a useful organisational form. It is difficult to see however any marked increase in co-operative research in view of the trend towards larger groupings in industry. Contract research on the other hand might provide one area of expansion, although some RAs are ill-equipped to undertake this. The financing of the RAs should be reviewed, and the possibilities for greater specificity in both industrial and government contributions examined.

NOTES on Chapter 8

1. For an historical survey see R. S. Edwards *Co-operative Industrial Research* London, Pitman (1956)
2. *Survey of Industrial Research in Manufacturing Industry* London, FBI (1960)
3. *The Structure of Industrial Research Associations* Paris, OECD (1965) p. 38
4. C. F. Carter and B. R. Williams *Industry and Technical Progress* Oxford, Oxford UP (1959) p. 217
5. See for example H. F. Heath and A. W. Hethington, *Industrial Research and Development* London, Faber (1964)
6. *Structure of Industrial Research Associations* op. cit. p. 37
7. *Grant Application for 1954–59* London, British Launderers' RA (1953)
8. D. McNeil 'Productivity in an Industrial Research Association Symposium on Productivity in Research' (Institution of Chemical Engineers, 11–12 December 1963)
9. This position is outlined by Dr W. Watson, Director of the Rubber and Plastics RA in *Rapra Jubilee* Shrewsbury RAPRA (1968)
10. Letter to RA directors from the Ministry of Technology
11. P. S. Johnson *Co-operative Research in Industry* London, Martin Robertson (1973)
12. Report of the FBI Conference on Industrial Research Eastbourne, 1962
13. P. S. Johnson op. cit. p. 164
14. P. S. Johnson 'The Role and Operation of Co-operative Research Associations in British Industry', Ph.D. thesis submitted to Nottingham University, 1970
15. *Research for Industry* London, HMSO 1961 pp. 37–41

9 The Universities' Contribution

This chapter considers the fourth major channel for scientific and technological work: the universities. As in previous chapters, the first section gives a brief outline of the importance of universities in this field, in so far as it can be ascertained from the available statistics. The second section is devoted to a discussion of the nature of the universities' output and of the problems of evaluating their activities.

9.1 The importance of the universities

As with other channels for scientific and technological work discussed earlier, the most easily available measure of the universities' role is in *input* terms, ie R & D expenditures. Table 6.1 on p. 116 shows the percentage of all R & D expenditure accounted for by the 'higher education sector' in the prinicipal OECD countries. The percentage varies between 8 and 34 per cent. These differences in turn reflect variations in the institutional framework, the structure and capabilities of industry and the R & D objectives of government. It is also worth noting from table 6.1 that 'other' sources of finance, which includes that from higher education itself and from foreign sources,* *finances* far less than the higher education sector spends. Government, and to a much lesser extent industry, are clearly important as sources of finance for R & D in the higher education sector. This arises from the fact that very little of

---

* Higher education is not isolated as a separate source of finance for R & D in the UNESCO statistics on which table 6.1 is based.

university research output can be 'sold' in the market. And because there is no direct market feedback to the universities, the allocation of resources to and within this sector poses particularly difficult problems. These problems are dealt with in greater detail later in the chapter.

In the UK the main sources of funds for the universities are given in table 9.1. (The figures also include finance for R & D conducted in other sectors of higher education, but this element is extremely small and can be ignored for the purposes of this chapter.) The government's contribution is clearly the largest at 82 per cent. (In the US the figure is about 75 per cent.[1]) It comes in different forms, as illustrated in table 9.2, although the two major contributors are

Table 9.1

Sources of finance for university scientific R & D in the UK, 1968–69

Source	£m.	%
Government	74.8*	82.1
Universities	6.8	7.5
Public corporations	0.6	0.7
Research associations	—	—
Private industry	3.6	4.0
Overseas	1.6	1.8
Other	3.7	4.1
Total	91.1	100.0

* Includes £2.9m. financed by local government.
Source: *Research and Development Expenditure* London, HMSO (1973) table 3C

Table 9.2

Gross government expenditure on university scientific R & D, 1970–71

Source	£m.	%
Defence	1.2	1.5
Research councils and other science grants	16.2	19.3
Trade and industry	1.2	1.4
Agriculture, fisheries and food	1.1	1.3
University Grants Committee	63.8	76.0
Other	0.5	0.6
Total	84.0	100.0

Source: *Research and Development Expenditure* London, HMSO (1973) table 3C

the University Grants Committee and the research councils. The former is a semi-independent body which allocates block grants to universities for both teaching and research. The universities are then free, within certain limits, to spend this money as they wish. As far as the finance of research is concerned, the work of the Committee was summarised in 1969 as giving a 'floor of support' for scientific research in universities : providing academic salaries, adequate supporting staff, library and computing facilities, basic and standard facilities of a research laboratory in a 'well-found laboratory' and some limited discretionary funds.[2] The work of the research councils has already been mentioned (chapter 6). The ways in which they provide financial support for university research vary. The main channels are : research grants for particular projects; support for research units; and maintenance fees for postgraduate students.

9.1.1 *Industrial support for university R & D*

Table 9.3 gives some indication of industrial support by industry for university R & D. Two features stand out from this table. Firstly, industrial support for university R & D is very small indeed : in only two product groups is it above three per cent of all R & D expenditure. This is to be expected bearing in mind the orientation of most university work towards the basic end of the R & D spectrum; even where a company is prepared to finance research of a more fundamental kind, it will usually wish to undertake it within its own organisation, where contact with applied research workers will be at its closest and where the maximum advantage can be taken of being 'first in the field'.

The clear difference in R & D orientation between industry and the universities is borne out by a recent survey carried out by a Working Party of the Universities and Industry Joint Committee. They found for example '. . . that most companies do not regard University deparments as a major source of technical advice except when they can offer specialised knowledge or expertise'.[3] The Working Party also investigated the nature of the obstacles to closer collaboration between industry and the universities. The obstacles most commonly mentioned by industry and the university departments was the difference in aims and outlook between the two sectors and difficulties in communication.[4] Universities do of course

Table 9.3
Industrial finance for university R & D

Product group	Finance for university R & D as % of total R & D expenditure in product group	% of total industrial finance for university R & D
Mining and quarrying	3.2	1.6
Food, drink and tobacco	0.5	5.9
Chemicals	0.6	34.4
Iron and steel	0.7	4.3
Non-ferrous metals	0.3	1.4
Mechanical engineering	0.3	8.7
Scientific instruments	0.2	2.2
Electrical engineering	0.2	16.5
Ships and marine engineering	0.4	1.2
Motor vehicles	0.1	1.8
Aerospace	less than 0.1	1.6
Railway equipment	3.3	1.6
Clothing and footwear	1.9	1.1
Pottery, china and glass	0.4	1.6
Rubber	0.5	1.3
Other manufacturing	0.1	2.2
Construction	4.3	8.2
Utilities and services	1.3	4.3
Total	0.3	100.0

Source: from *Research and Development Expenditure* London, HMSO (1973) table 15D

vary in the extent to which their research is geared to industrial needs. For example it is probably true to say that the research work of the technological universities created in 1965 from the colleges of advanced technology is more orientated towards industry than that of most other universities.[5]

The second noteworthy feature about table 9.3 is that sources of industrial finance for universities are heavily concentrated in a few industries: chemicals, engineering and construction account for nearly 68 per cent of the total. This concentration is hardly surprising since the academic subjects that relate to those industries are among the most well established in universities.

Support for research in universities tends to come from the larger firm: the study by the Universities and Industry Joint Committee referred to above showed that a far higher proportion of companies of over 5,000 employees sponsored university work than did smaller sized companies.[6] They also had relatively more contact with universities in other ways (eg through seminars, lectures etc.)

9.1.2 *The nature of the universities' research*

Practically all university research is on basic and applied research rather than on development.[7] It is also probably true to say that the emphasis of most of the work classified as 'applied research' is orientated towards the basic rather than the development end of the spectrum, although the problems of definition are particularly acute here and are reflected in the official statistics: for example in 1964–65, the latter estimated that about 90 per cent of all university research was of a basic nature; between 1966–7 and 1968–9 a percentage of around 50 was given; and in 1969–70, no breakdown was published!

9.2 Evaluating university research

Because resources allocated to research have an opportunity cost, some criteria and mechanisms for allocating resources between R & D and other unrelated activities, between the different types of research and development and between projects within each type, must therefore be established. How much university research should be financed? How should this finance be allocated between projects? In which universities should the research be carried out? If some measure of university research *output,* valued in money terms, could be established then the allocation problem would be greatly eased. Even measurement without valuation would be a useful first step since it would enable some limited cost effectiveness work to be undertaken (although it would not answer the basic question of whether a project was worthwhile). As we shall see however university research output presents very considerable measurement problems.

Firstly, even *defining* such output raises fundamental issues related to the nature and purposes of universities. It would be generally accepted however that it is of a multi-dimensional character (see below). Consequently, without some weighting system it would be impossible to compare projects. Such weighting clearly involves value judgements. Secondly, university research output is often so long-term and its contribution so indirect that if measurement of output is to assist decision-taking, it must be made on an *ex ante* basis. But by the very nature of much university research work, accurate prediction of the results of a particular programme or project is often impossible.

Nevertheless, the existence of these difficulties does not alter the fact that the resources still have to be allocated on some basis. Useful insights into the kind of problem that has to be faced here may be obtained by analysing what are generally considered to be the main dimensions of university research output. These dimensions may be classified under three main headings. Firstly, university research acts as a means for improving the stock of scientific manpower. It does this in a number of ways. It provides a training ground for postgraduate students, giving them experience of research techniques and a detailed knowledge of the latest developments in a particular area. This training role is important because not only does it train manpower to conduct future research, but it also provides an appreciation of the research that is being carried on elsewhere. Research is also likely to affect the teaching of undergraduates. The relationship between research and teaching has long been a source of controversy : on the one hand, it is frequently argued that research activity improves the quality of a lecturer's teaching; on the other, research is said to divert resources away from teaching. Unfortunately, little work has been done on the relationship between these two elements although the dual nature of universities has been an important element of higher education policy in most advanced countries.

Secondly, university research output takes the more obvious form of research results. The nature of these results will vary considerably in their economic implications. Some may have a fairly immediate impact in new or improved products' end processes. Professor J. F. Baker's work on plastic flow at Cambridge, for instance, was taken up fairly quickly in construction design.[8] Some may lead to new research techniques and equipment that are then utilised in later research. For example, the cyclotron was first conceived and designed in a university.[9] Some may have an impact on the medical sphere, which in turn may filter through to increased longevity etc. Fleming's discovery of penicillin while at the University of London is perhaps one of the most dramatic examples of this nature.[10] Most university research however will simply increase knowledge in a particular field and may have a very limited impact in the short run. However in the longer run – two, three or even four generations later – this knowledge may have far-reaching effects, contributing perhaps to entirely new industries. Much university research will first become absorbed into the 'scientific consensus' before leading to new technology. A study conducted in the United States in the 1960s which was concerned, *inter alia,* with the role played by basic research in the development of weapons systems in the postwar period, led the investigators to suspect that, however important scientific advance may be, 'its primary impact may be brought to bear not so much through the recent, random scraps of new knowledge, but through the organised, 'packed down', thoroughly understood, and carefully taught *old* science'.[11] This is not to say however that major and fairly discrete advances in scientific knowledge do not occur. For example the work by Maxwell, Thomson and Rutherford at the Cavendish Laboratory, Cambridge University, is clearly identifiable as representing notable advances in the understanding of electromagnetism, the electron and the atom respectively.[12]

Nevertheless the *mechanism* by which new knowledge becomes incorporated into technology may not be as simple as might at first be thought. The idea of a scientific discovery leading directly to inventions which lead in turn to innovation is a gross oversimplification for two reasons. Firstly, a new product often results from a whole series of interacting discoveries and inventions that extend over a long period of time. For example, while the key basic concepts underlying the transistor emerged in the 1940s, their antecedents can be traced through a series of discoveries stretching back into the nineteenth century.[13] Secondly, it is by no means certain that the invariable causal flow is from scientific discovery to invention. While it is probably true to say that, in the majority of cases, basic research precedes applied research which in turn precedes invention, it is by no means universally true.[14] An invention may provide the stimulus for more basic research to be undertaken. For example, the invention by Cockerell of the hovercraft acted as a stimulus for basic work on the behaviour of the air cushion under different conditions (see part 3). Even where the basic research precedes (in time) the applied research, this does not necessarily imply anything about a causal relationship. A particular invention may lead to the discovery of existing work which has remained unknown. Thirdly, an important discovery may arise almost by accident, rather than from a closely structured research project with rigorously defined objectives. For example, polythene was discovered as a result of a fault in some laboratory apparatus.[15]

In so far as it is possible to identify discrete results of basic research, it is likely that by far the biggest share of these results comes from the universities. Some limited quantitative evidence on this score in the United States is provided by a study carried out by the Illinois Institute of Technology Research which undertook the retrospective tracing of key events that had led to five important technological innovations. (The use of the word 'led' implies some causality, although there is little direct evidence in the study to support this.) The choice of key events inevitably involved some subjective assessment, although 'general agreement among the scientists involved on what constitutes an important event was reached fairly easily'.[16] Key events were then classified into three classes: non-mission research; mission-orientated research; and development and application. These definitions compare broadly with the division between basic research, applied research and development given on pp. 20–1. The investigators found the breakdown of events by performer as given in table 9.4. Clearly the universities (both US and foreign) were the major source of non-mission research events, but they were much less important in the two other categories.

Table 9.4
Distribution of key events by performer, five innovations

	Universities & colleges	Research institutions and governmental laboratories	Industry
Non-mission research	76%	14%	10%
Mission-orientated research	31%	15%	54%
Development and application	7%	10%	83%

Source: *Technology in Retrospect and Critical Events in Science* Washington, National Science Foundation (1968)

However it is also true (as the table shows) that research work in industry has led to some fundamental advances. An important example here is the semiconductor research at Bell Telephone Laboratories in the late 1940s, which led to the transistor (three of the investigators shared a Nobel prize). This example is especially interesting as it illustrates some of the problems of defining 'basic' research. Bell believed that 'major advances in scientific knowledge in this field were likely to be won and that advances in knowledge were likely to be fruitful in improving communications technology'.[17] The work they sponsored however was clearly of a very theoretical nature and several of the scientists involved 'were not much interested in or concerned with any practical applications their work might lead to'. Yet it seems fairly clear that Bell would not have supported the work if it did not think that there could be some economic pay-off in the long run.

The third aspect of university research output is its 'cultural' value, which arises from the greater knowledge that it provides about man and his environment.

It is clear from the above discussion that the output from university research is very complex and often not readily identifiable even in *ex post* terms. In consequence the economic evaluation of expenditure in this area is likely to meet with very formidable measurement difficulties. The most difficult problems of evaluation arise in the basic research field, and the rest of this section will concentrate on the problems in this area.

One very exploratory attempt to bring economic analysis to bear in this field has been made by Byatt and Cohen.[18] In their effort to provide an evaluation mechanism, they start off from the premise that basic research often gives rise, after a substantial time lag and with other inputs, to new industries or to new sectors in established industries. From the establishment of these inputs a stream of net benefits is generated which can be discounted to a suitable starting date (for example, the date when a discovery starts to become economically relevant, ie when applied research begins). Byatt and Cohen then suggest that one way of estimating the value of a scientific discovery is to analyse the effects, in terms of net benefits, of a change in the time at which the discovery was made. This change would in turn result from an alteration in research inputs. The approach they suggest would only be of value therefore in the assessment of *marginal* changes in research inputs and not for evaluating whole sectors of research. However the former is likely to be of greatest use in policy terms as piecemeal adjustments rather than wholesale changes are likely to be the most common implication of policy decisions.

Unfortunately there are at least two major problems in the Byatt and Cohen approach which, if it is eventually to be of use, must be applicable on an *ex ante* basis. (We ignore here problems associated with cost–benefit analysis in general, eg the choice of a discount rate; we shall also ignore for the moment the fact that Byatt and Cohen assume a causal relationship between scientific discovery and industrial activity.) Firstly, tracing the changes in benefits that may result from a change in the timing of a particular discovery may be very difficult because of the complex interaction of different discoveries. In some cases such a change may make very little difference because of the already considerable lag between the discovery and the inception of industrial activity. Byatt and Cohen use the example of Faraday's work to illustrate this difficulty: since his discovery of electromagnetic induction in 1831 did not have any important economic pay-off until 1875–80, '[a] ten year delay in the discovery would probably not have resulted in a ten year delay in the use of dynamos in commercial electric lighting'.[19] Secondly, for the method to be useful it is necessary to establish a link between research inputs, ie expenditure, and output in the form of discoveries. Unless this can be done there is no way in which the economic benefits of curtailing or expanding a research programme can be estimated. By its very nature, however, the output from a substantial amount of basic research is unpredictable; nevertheless it is worth remembering, as Foster has pointed out, that we do not *behave* as if successful research was an entirely random event: we attempt to ensure that finance is going to the 'right' man.[20] It may be possible therefore for scientists to give *some* indication of the broad nature of the likely results although this is rarely likely to be in a precise form.

The two problems outlined above are further complicated by the international nature of basic research finidings (see pp. 23–4). New industries founded in the UK may be based, in part or in whole, on discoveries made overseas. Thus to estimate the effects of a change in UK basic research expenditure on future UK industrial development must also involve an analysis of what would have happened in other countries. If for example several countries (including the UK) with roughly similar resources were working in the same area, a reduction in UK basic research may not affect the outcome except in so far as it may reduce the ability of the UK either to appreciate results obtained elsewhere, or to gain some commercial advantage resulting from its being first in the field.

The Council for Scientific Policy which sponsored Byatt and

Cohen's work, took up their suggestion that some exploratory *ex post* studies should be undertaken to examine the feasibility of their approach.[21] The Department of Liberal Studies in Science at Manchester was asked to look at two recent major innovations, the Chorleywood Bread Process, (CBP) and Float Glass. The first, which was introduced by a research association (see p. 199), involves the replacement of a previous lengthy period of dough development by a much shorter mechanical and chemical treatment. The investigators tried to trace what contribution, if any, from 'curiosity-orientated' research had been important in its development. They were however '... forced to the conclusion that there is no identifiable curiosity-oriented research of the present century which, had it been delayed, would have occasioned a corresponding delay in the development and adoption of the CBP'.[22]

The study of float glass showed that much of the scientific work on the process occurred *after* it had been developed empirically. Furthermore, although much of this scientific work utilised some fundamental concepts and analytical techniques which may have had their origin in curiosity-orientated research, '... it does not seem possible to apply the concept of marginal delay to these science–technology connections...'.[23]

There are of course cases where scientific discovery has given rise fairly directly to innovations. Kipping's work on organo-silicon chemistry at University College, Nottingham (as it was then), is clearly identifiable as the basis from which silicones were developed commercially, although Kipping himself was not interested in industrial applications.[24] Even here however the relationship between scientific discovery and commercial use is obscured by the intervention of the First World War.[25] The time period between these two elements is usually so long that it will be rare for it to be clear of 'special' factors of this kind.

If the Byatt and Cohen method cannot be used to allocate resources to basic research on the grounds that both its assumptions and feasibility are doubtful, a less ambitious approach must be adopted. Nearly all basic research funds in this country are allocated on the basis of a qualitative assessment of the field or project – usually by a committee. But what criteria should be used as guidelines? Michael Polanyi has argued that scientists should be able to allocate the funds themselves on the basis of scientific excellence alone: an 'invisible hand' (a term familiar to all economists) would guide resources into those areas with the greatest scientific potential.[26] In this scheme of things, science is pursued by 'a mass of independent self co-ordinated initiatives'. The Council for Scientific Policy has however criticised this viewpoint, arguing that 'science is not a self-contained and self-determining activity insulated from social, political and economic activities'.[27] The Council has instead constructed its own 'shopping list' of criteria, which includes not only the 'intrinsic' factors that it thought Polanyi was overstressing, but also 'external' and 'resource implications' factors.[28] External criteria cover such factors as the short- and long-run economic benefits of the proposal, and its effects on national prestige and reputation. 'Resource implications' relate to the project's effect on the demand for manpower and capital.

Two points may be made on this list of criteria. Firstly, it is by no means clear that 'intrinsic' and 'external' criteria can be distinguished at all easily: scientific 'excellence' or 'potential' may imply that some of the latter criteria are satisfied. Secondly, while the checklist may have some use in decision-taking, the basic problem of weighting the different elements remains.

9.2.1 *Measuring productivity in university research*

It has already been shown that a rigorous economic assessment of university research is not possible; however, it may be possible to construct a series of indices which attempt to quantify the output of particular research workers or groups of research workers. This must necessarily be done on an *ex post* basis, but such an exercise may be of value in giving decision-takers some – albeit imperfect – guidance on where given financial support may lead to the greatest output. It may also assist in answering the question of whether concentration of inputs, for example in research units, raises the productivity of research workers.

As we have seen, the measurement of university research output in physical terms does of course require some consensus about what university research is for. Furthermore, where this output has to be measured in several dimensions, some weighting of the factors involved is necessary. An attempt to measure the research output of university staff in chemistry has been made by Blume and Sinclair.[29] They used the following five output indicators:

1) *publications,* as measured by a simple count of papers published

and awaiting publication over the previous five years (no weighting by importance was attempted; there is some evidence to suggest that the results obtained from a weighting system would not have been very different from those provided by a simple count);[30]

2) *doctorates awarded to students* over the past five years;

3) *relevance,* as measured by the number of Co-operative Awards in Pure Science awards obtained, the number of patents held, and the number of days spent on consultancy in the previous year;

4) *peer group judgement* : (a) *recognition,* as measured by honours and medals received, 'honorific' functions performed (eg membership of the editorial board of an academic journal) expenses-paid invitations from overseas and the number of senior visitors received by the research group;

5) *peer group judgement* : (b) *assessment,* for which respondents were asked to name individuals whom they considered to be pacemakers in their particular field, both within the UK and overseas (each respondent was then given a score based on the votes received).

Data on these output measures were obtained from a questionnaire sent out to all those undertaking university research in chemistry, except students. Where the measures involved more than one indicator these were combined using a points system constructed by the investigators. This does of course mean that their own weighting values were used.

One particularly interesting outcome of this study was that the different measures of research output outlined above do not appear to be highly correlated with each other, as table 9.5 shows. In particular, it is worth noting that industrial relevance does not appear to be closely correlated with either recognition or papers published. Blume and Sinclair suggest that a tentative conclusion from this data 'might be that those whose work has been the most closely geared to industrial interests have been neither the most prolific . . . nor the most esteemed'.[31]

9.2.2 *The concentration of resources*

In recent years there has been a trend towards the concentration of research resources in particular universities and research units. The arguments for such concentration are principally threefold.[32] Firstly, it avoids duplication of research effort. Secondly, 'it enables researchers to work closely with colleagues who are working in the

Table 9.5
*Relationships between various measures of research output,
British university chemistry research*

Papers published	Papers published			
PhDs supervised	0.51	PhDs supervised		
Industrial relevance	0.15	0.21	Industrial relevance	
Recognition score	0.44	0.48	0.13	Recognition score
Assessment score	(0.65)*	—	(0.44)*	(0.76)* Assessment score

* Figures in brackets are γs (the Goodman–Kruskal γ); others are Spearman value correlation coefficients and the two are not strictly comparable. For an explanation of the first term, see R. S. Weiss *Statistics in Social Research* New York, Wiley (1968) chapter 11.

Source: as n. 29, p. 22

same or allied disciplines. This then creates a synergic effect. Thirdly, because some equipment is so expensive, a 'threshold' level of expenditure often exists which must be reached before a viable team can be supported. This may be particularly true in the field of 'big science' research, ie research in nuclear physics, radio astronomy and space research. The principal arguments against concentration are: 1) that it does away with competing centres of research; if the competitive spirit is one element in research motivation, concentration may serve to *lower* output; 2) that in universities that are not endowed with such centres, teaching may suffer in consequence; and 3) that it may create a scientific bureaucracy that stifles dissent and originality.

Rightly or wrongly, greater concentration has been pursued by the Science Research Council. In 1970, its policy was summed up as follows:

> Certain areas within a discipline or embracing a number of disciplines will be selected for more favourable than average support during a given period, on the basis of a review of their special potential for advancing basic science, or their economic or community value, or all three . . . This concentration of resources will be planned by shifting to favoured areas from less favoured areas, rather than by simple addition.[33]

The evidence on whether larger units are in fact more productive is extremely sparse. Blume and Sinclair found in their study of university chemistry research that, although there was a statistically significant relationship between the size of the research groups within which respondents worked and the research output measures they used, it was not a strong one.[34] They also found no evidence to support the idea of a 'threshold' level. However they did find that the strength of the relationship between size and performance varied systematically with the academic rank of the respondent, and was strongest for readers, who are the most committed to scientific research. This may suggest that 'a large research group is particularly beneficial to the scientist who is himself deeply involved in its research'.[35] Furthermore, the correlations also varied systematically between subdisciplines of chemistry. Even if the correlation coefficients were very high for all ranks and subdisiplines, however, it would not prove that large research groups are more effective than smaller ones; it may merely reflect the fact that eminent research workers are usually able to attract relatively more co-workers. In interpreting Blume and Sinclair's results it must be remembered that their work was limited to chemistry, where the threshold level may be much lower than in the 'big science' subjects.

The discussion above has been in terms of research group size and *output*; but what of the *costs* of running different sized research groups? Again the evidence is very fragmentary. As far as university *departments* are concerned, it appears that *per capita* equipment costs *rise* as a department becomes larger.[36] The implications of this finding are not however unambiguous, since it may reflect not so much greater inefficiency as size increases but the fact that larger departments tend to attract eminent scientists as their heads and that in consequence, these departments tend to attract more research funds for the latest equipment.

9.3 Conclusion

Although the universities in the UK spend around £100m. on research in this country, and although in the long run their research activities are probably of very considerable importance in stimulating technological advance and in increasing its rate, the economist, for the present at least, can be of only limited assistance in the allocation of resources to or within this sector. This impotence is due principally to the complex and unpredictable nature of university research output and to the fact that the latter is often apparent only in the very long run. Attempts to provide a more rigorous economic framework for decision-taking have so far been unsuccessful. It seems inevitable therefore that most of the resource decisions must still be taken on the grounds of qualitative assessment. However this does not mean that those aspects of basic research that can be quantified should not be so, or that basic economic issues should not be made explicit in the qualitative assessment.

Some attempt has been made to measure university research output in physical terms. A most interesting result of this work is that the industrial relevance of the 'output' has not been closely related to the esteem in which the research worker is held by his professional colleagues. It has also been shown that there is room for doubt, in the field of chemistry at least, whether research group size is strongly positively correlated with research productivity. On the *costs* side the evidence suggests that, for science and engineering departments, *per capita* equipment costs rise with the size of department. This may not however imply inefficiency but rather that bigger departments have better reputations and are therefore better endowed in equipment terms.

NOTES on Chapter 9

1. R. R. Nelson et al. *Technology, Economic Growth and Public Policy* Washington, Brookings Institution (1967) p. 59
2. Council for Scientific Policy *Report of a Study on the Support of Scientific Research in the Universities* (Cmnd. 4798) London, HMSO (1971) p. 17
3. *Industry, Science and the Universities* (Report of a Working Party on Universities and Industrial Research to the Universities and Industry Joint Committee) London, CBI (1970) p. 49
4. *ibid.* pp. 56–7, 88
5. M. Sanderson *The Universities and British Industry* London, Routledge & Kegan Paul (1972) chapter 13
6. *ibid.* p. 35
7. *Research and Development Expenditure* 1970 London, HMSO (1973) table 6
8. J. Langrish et al. *Wealth from Knowledge* London, Macmillan (1972) pp. 409–16
9. J. Jewkes et al. *The Sources of Invention* 2nd edn, London, Macmillan (1969) pp. 248–9
10. *ibid.* pp. 278–9
11. C. W. Sherwin and Raymond S. Isenson 'Project Hindsight' *Science* (23 June 1967)

THE UNIVERSITIES' CONTRIBUTION   225

12. A. E. E. Mackenzie *The Major Achievements of Science,* Cambridge, Cambridge UP (1960) chapter 20
13. I. C. R. Byatt and A. V. Cohen *An Attempt to Quantify the Economic Benefits of Scientific Research* (Science Policy Studies No. 4) London, HMSO (1969) p. 11
14. *Technology in Retrospect and Critical Events in Science* Washington, National Science Foundation (1968) Part III
15. J. A. Allen *Studies in Innovation in the Steel and Chemical Industries* Manchester, Manchester UP (1967) p. 21
16. *Technology in Retrospect and Critical Events in Science,* Washington, National Science Foundation (1968) Part III, p. xiii
17. R. R. Nelson 'The Link between Science and Invention' National Bureau of Economic Research *The Rate and Direction of Inventive Activity* Princeton, Princeton UP (1965) p. 560
18. I. C. R. Byatt and A. V. Cohen op. cit.
19. ibid. p. 13
20. C. D. Foster 'Cost Benefit Analysis in Research' in A. De Reuck *et al.* (eds) *Decision Making in National Science Policy* London, CIBA Foundation (1968)
21. *Third Report of the Council for Scientific Policy* (Cmnd. 5117) London, HMSO (1972)
22. J. Langrish *et al.* op. cit. p. 148
23. ibid. p. 38
24. J. Jewkes *et al.* pp. 296-8
25. J. Langrish *et al.* op. cit. p. 39
26. M. Polanyi 'The Republic of Science' *Minerva* (1962-63)
27. *Third Report of the Council for Scientific Policy* (Cmnd. 5117) London, HMSO (1972) p. 19
28. ibid. pp. 20-1
29. S. S. Blume and R. Sinclair *Research Environment and Performance in British University Chemistry* (Science Policy Studies No. 6) London, HMSO (1973)
30. J. and S. Cole 'Scientific Output and Recognition: A study of the Operation of the Reward System in Science' *American Sociological Review* No. 32 (1967) quoted in Blume and Sinclair op. cit.
31. S. S. Blume and R. Sinclair op. cit. p. 22
32. ibid.
33. *Report of a Study on the Support of Scientific Research in the Universities* op. cit. p. 53
34. S. S. Blume and R. Sinclair op. cit. Section IV
35. ibid. p. 50
36. E. G. Bevan *An Analysis of Equipment Costs in University Science and Engineering Departments* (Science Policy Studies No. 35) London, HMSO (1967)

PART 3

A Case Study in Invention and Innovation: the Development of the Hovercraft

10 The Background to the Study

This part analyses the economic and technical development of the hovercraft and associated craft in the UK, in the light of the discussion in the preceding parts. Although a number of inventors and designers have experimented at various intervals with the idea of using an air cushion in various forms of transport and equipment for many years (see chapter 13 below), it is generally accepted that modern hovercraft development effectively starts from the mid-1950s. It is also usually accepted that the British inventor, Christopher Cockerell (now Sir), has the 'earliest priority date' for the key patents, although several people were working along broadly similar lines at roughly the same time.

Because the development of hovercraft is fairly recent, the author has had the advantage of meeting many of the people who were concerned with the beginnings of its technical and commercial exploitation. Most of these people are still working in or connected with the industry and hence are able to provide a useful perspective on early events and decisions. The author has also interviewed businessmen and others who have become involved with hovercraft (on both the production and the operating side) more recently. The industry has not yet settled down into any kind of stability; consequently it has been possible to experience at close quarters something of the excitement and problems involved in an innovating process. Companies that have recently commenced or withdrawn from hovercraft operation and production, companies that are facing or have faced major financial problems and companies that are beginning to show signs of success have all been visited.

However, the fact that developments over the past sixteen or so years are being dealt with brings some disadvantages. Firstly, it is difficult to put the role played by different individuals, companies and institutions into perspective. This problem arises in part because it is not yet known whether the hovercraft, from an *ex ante* viewpoint, can be classed as a successful or unsuccessful innovation in economic terms, despite the enthusiasm of Jewkes *et al.*[1] 'Success' can of course be defined in different ways — for example in social or in private terms. If the former yardstick is used, then the problem is conceptually relatively simple : the project should show a positive net present social value, or alternatively a social rate of return at least as good as that obtainable elsewhere. However, as we have seen, the calculations necessary for a social cost–benefit analysis raise a number of thorny problems of valuation, especially in relation to benefits. The author has not attempted a full cost–benefit study of hovercraft development although some idea of a few of the magnitudes (in relation to marine hovercraft) may be given for the period up to the end of 1972. The principal, although not necessarily the only, benefit on the civil side is the passenger time savings on commercial ferries made possible by the higher speed of hovercraft over conventional marine transport. If the calculations and assumptions in appendix 1 are accepted, the value of total time savings up to 1972, discounted to 1959, the effective start of hovercraft development, are probably in the region of £1m. if a 5 per cent discount rate is used (£1½m. with a 10 per cent rate). (It is the author's view that the assumptions made are probably favourable overall to hovercraft.) Other benefits could include the convenience of hovercraft travel (reduced queues at terminals, greater frequency, etc), but these are likely to be marginal. On the military side, some benefits may have been derived from hovercraft operation, but they would be difficult to quantify. To date, the total benefits in this sphere are probably very small and relate principally to specific operating characteristics of hovercraft rather than to any general advantage they have over existing forms of transport (see below).

On the cost side, appendix 2 suggests that total (including operating) costs to 1972 net of foreign sales, may be of the order of £26m. in 1959 terms using a 10 per cent discount rate, £16m. with a 15 per cent rate. (Higher rates are used to discount costs to allow for the effects of inflation. When calculating the benefits from time savings, the actual time saved in hours for each year was estimated and then discounted to 1959. It was then valued in 1959 terms. Hence no allowance for inflation was necessary.) From these estimates, any cost savings resulting from reduced investment in and operation of ships, and the returns from releasing existing ships for other uses, have to be subtracted. It is doubtful whether these offsetting factors would amount to more than £5–6m. as at 1959 (using a 10 per cent discount rate). Thus at the end of 1972 the development of hovercraft was in substantial deficit even when allowance is made for the very tentative nature of the figures.

This result may be partly as expected: forecasts of *future* costs and returns would of course be crucial to a full cost–benefit appraisal. Clearly, part of the total expenditure figures used above represents investment that has not yet received any returns. Furthermore, there are numerous examples of inventions in both transport (eg aviation)[2] and other industries (eg radio)[3] that had very long 'gestation' periods before they became profitable. Enos's data (p. 174) show that an extensive interval may exist between the date of an invention and its eventual commercial introduction. And the latter event will usually occur years before the innovation becomes profitable. Nevertheless, for the moment we must remain at the very least agnostic about the ultimate success of hovercraft development in *ex ante* social terms: it would take a very substantial increase in the growth of the commercial market and a very substantial decrease in manufacturing and operating costs to make the account profitable in this way. The scale of the problem may be illustrated from the fact that the 1959 value of 1972 operating costs *alone* (excluding depreciation) was probably over £600 thousand (using a 15 per cent discount rate); the total time savings (discounted at 10 per cent) given the author's values, amounted to around £125 thousand. It is the author's view that if the overall social account is eventually to show a profit, that success will not come principally from the commercial operation of amphibious craft in this country. In a recent cost–benefit study of the Channel tunnel, independent consultants to the government stated:

> The present (cross channel) hovercraft services ... do not appear to offer a satisfactory alternative to the ferries. Operating costs are considerably higher per capacity mile and even with relatively high load factors the present services are either operating at a loss or are at best marginally profitable. By contrast, the existing shipping services on the equivalent routes earn very satisfactory profits even at relatively low load factors. Consequently there

seems little future for the hovercraft as a majority carrier although without the Tunnel a restricted service could possibly cater for a limited segment of the market prepared to pay a substantial tariff premium to obtain a reduction in journey time.

We have discussed the prospects for further development of hovercraft services with officials at the Department of Trade and Industry. There is inevitably some uncertainty on this but the Department's view was that there appears to be little potential for further development of hovercraft services unless there are substantial and as yet unforeseen developments in hovercraft technology.[4]*

This statement referred only to the BHC 190–ton (250 passengers, 30 cars) SRN4 which is the only type of hovercraft used on this route.†

Operating costs on the smaller (55 passenger) SRN6 are relatively high.‡ In 1971 Hovertravel estimated that its break-even point for operating costs (at 60 per cent pay load) was about 12p per seat mile on this craft. This compared with fares reflecting seat mile charges of around 4p on conventional water transport.[5]

The consultants' view outlined above is consistent with the policy now adopted by both BHC§ and Vosper Thornycroft of aiming primarily at military markets, where cost factors are less important, and with the fact that the smaller amphibious craft now being produced are not intended primarily for commercial ferries (see chapter 11). It must be remembered, however, that the consultants' judgement was based on experience with *amphibious* rather than with sidewall craft.

If the private yardstick of success is used, a problem arises because some companies may be very successful while others may fail. In this case one can only talk of experience of individual companies. Many companies are still in the development or early commercial stage; whether their craft will sell and be profitable is still unknown. Although in the last few years some of the companies involved

---

* The Department's rather pessimistic view on developments in hovercraft technology seems to be consistent with BHC's current attitude to hovercraft (see p. 240). They may not of course be independent!
† The SRN4s have since been 'stretched' to carry 280 passengers and 37 cars.
‡ Only SRN4s are used on the cross-Channel service to which the quotation in the text refers. SRN6s are employed principally on the Solent.
§ BHC was formed in 1966 following the amalgamation of the hovercraft interests of Westland Aircraft and Vickers.

have made a profit, sometimes substantial (see chapter 14 below) most are far from achieving an adequate and/or stable return on their hovercraft investment. Vickers and Denny Hovercraft both abandoned their hovercraft work without achieving any sales. Vosper Thornycroft has still not sold any craft, six years after entering the field. Up to 1972 neither of the two big operating companies had made a profit (see p. 271). Once more however, it must be remembered that initial losses on important inventions have often been high.

The above discussion should not be taken to imply that it would be non-economic either for the country as a whole or for individual firms to continue in hovercraft production or operation, even if there were no prospect of ever earning an adequate return on the *total* investment, since at any given point of time costs incurred in the past and now irretrievable are irrelevant to current decision-taking. Looking at hovercraft development *ex ante*, however, an adequate social or private return has not yet been received. The NRDC has to date also made a considerable loss on its hovercraft activities (see chapter 15).

Knowing whether a development is successful or not is clearly important in making an assessment of the organisations involved. For example, in looking at the characteristics of those companies that have entered or refused to enter the hovercraft field we do not always know for certain at this stage whether companies who make right or wrong decisions are being examined, although we may make our own judgements. This problem should resolve itself in time.

The second difficulty arising from conducting a study at this relatively early stage arises from the reluctance of organisations and individuals to divulge material on such recent events. On the industrial side, this problem is compounded by the fact that the British scene is dominated by one company, BHC. Aggregation of data therefore would be unlikely to disguise BHC's position. On the public side, NRDC formally refused to co-operate in providing information although considerable help was given by individuals on a personal basis. While the Department of (Trade and) Industry was very willing to help in general terms, it did not have, or was unwilling to provide, statistical material on its hovercraft activities. The non-availability of information on public investment in hovercraft has inevitably severely restricted the scope of this study. One may also be a little reticent in assessing the part played by individuals still in the industry, yet their roles and personal circumstances may be crucial. As Sturmey says in his study of the development of radio, 'Alongside of all economic forces and motives . . . are the personal factors . . . If the economist ignores these personal factors he risks making his generalisations sterile and academic'.[6]

In the author's view however these problems are outweighed by the advantages of being near in time to the invention and its early development. Furthermore, major participation by NRDC in hovercraft development has now ceased and the government has said that it would give no further special aid to the industry after 1974 (see p. 307). Consequently one phase, ie that of close public involvement in hovercraft development, has now terminated. In any case, it is hoped to continue monitoring the development of the industry in the future.

An economic study of hovercraft development is attractive because its technical exploitation has involved many features of the invention and innovation process that are of particular interest to economists and are subject to some controversy. Generalisation on these topics cannot of course be made on the basis of one case study. However, studies of individual inventions and innovations are often a good seed-bed for hypotheses that may have a more generalised application.

Because of limited research resources, the study has been restricted to the development and exploitation of the air cushion principle in this country. The omission of overseas development, especially in France and the United States, makes the study in some respects incomplete, since technical competition from these countries has almost certainly influenced the policies of the British government and those of individual companies. However it is hoped to rectify this omission at some later stage. The study is principally concerned with marine and industrial applications. Reference is however made to hovertrain development in chapter 12.

NOTES on chapter 10

1. Jewkes *et al. The Sources of Invention* 2nd edn, London, Macmillan (1968) pp. 329–32

THE BACKGROUND TO THE STUDY 235

2. H. J. Dyos and D. H. Aldcroft *British Transport* Leicester UP (1969) chap. 13
3. S. G. Sturmey *The Economic Development of Radio* London, Duckworth (1958) p. 82
4. *The Channel Tunnel: A United Kingdom Transport Cost Benefit Study* (A Report by Coopers and Lybrand Associates Ltd) London, HMSO (1973) p. 19
5. E. W. H. Griffiths 'Current User Experience' *Production Engineer* (1971)
6. S. G. Sturmey op. cit. p. 277

11 Hovercraft Production, Operations and Markets

11.1 Production and operations

Turnover in the industry from both production and operations between 1967 and 1972 is given in table 11.1. (1967 was the first full year of BHC's existence. Prior to this year hovercraft work was undertaken by divisions of other companies for which separate sales figures are not available – see chapter 14.) BHC's turnover

Table 11.1
Hovercraft production and operations: turnover
(£000s)

Year	BHC	Producers*			Operators*		British Rail Hovercraft
		Hover-marine Transport	Cushion-craft	Air Cushion Equipment	Hover-lloyd	Hover-travel	
1967	1603		8		59		101
1968	5794		14		47	169	215
1969	10713		1		1068	273	636
1970	5336	173	121	18	1771	377	1273
1971	13354	505	—	86	2686	407	2253
1972	10186	841	—	110	3500†	241	2503

* Only companies that have had a turnover in excess of £100,000 p.a. some time in the period 1967–72 are included.
† estimated
Source: Company reports

includes revenue from non-hovercraft sources, eg from its production of Islander aircraft for Britten–Norman. About two-thirds of BHC's turnover in 1972 was directly attributable to hovercraft activities.[1] Percentages for earlier years are not available.

It is clear that BHC is almost synonymous with the British

236

hovercraft industry. There are several other companies actively involved in marine hovercraft development and production – principally, Air Bearings, Air Vehicles, Hovermarine Transport, Sealand Hovercraft and Vosper Thornycroft. At present (1974) however only Hovermarine Transport and Sealand Hovercraft have obtained any commercial sales. Air Cushion Equipment mentioned in the table utilises the air cushion principle for industrial applications (see chapter 14 below). It can be seen that total turnover in production has been very erratic. This reflects the 'bunching' of large orders for BHC in particular years. For example, the 1971 figures reflect large orders from Iran and Saudi Arabia.

On the operating side, the existing industry is dominated, in turnover terms by Hoverlloyd and British Rail Hovercraft ('Seaspeed'). Both operate cross-Channel services and BR Hovercraft also runs services on the Solent. The other company, Hovertravel, operates primarily on the Solent. The number of passengers carried by these operators up to 1972 is given in table 11.2.

Table 11.2
Passengers and cars carried on hovercraft
(000s)

Year	BR Hovercraft (cars in brackets)	Hoverlloyd (cars in brackets)	Hovertravel	Equivalent passenger miles (000s)
1965			180	720
1966			349	1850
1967	140		286	3267
1968	298 (3)		341	4693
1969	544 (18)	300 (40)	377	12,996
1970	682 (54)	500 (65)	352	25,824
1971	1034 (98)	618 (89)	389	35,511
1972	848 (103)	684 (103)	400*	36,663

* estimated

Sources: BR Hovercraft, Hoverlloyd, Hovertravel

As the journeys involved are of different lengths, the equivalent passenger miles have also been estimated. The latter data provide an S-curve of customer acceptance, the most rapid rise occurring between 1968 and 1970 while the cross-Channel services were becoming established. It remains to be seen whether the considerable levelling off in 1971 and 1972 will continue.

The share of hovercraft in cross-Channel sea traffic in 1971 is

238 CASE STUDY: DEVELOPMENT OF THE HOVERCRAFT

given in table 11.3. Hovercraft have clearly made substantial inroads into total traffic in the French Straits achieving well over 25 per cent within four years of operation. These gains must however be seen in the light of the losses made by the operators concerned (see table 14.6), and of the high fare structures existing on conventional shipping ferries on the Channel (see p. 295).

Table 11.3
Share of hovercraft in cross-Channel traffic, 1971

	Accompanied passengers*	Unaccompanied passengers	Total passengers*	Car equivalents†
% of total cross-Channel sea traffic	13.7	12.8	13.2	12.6
% of total traffic in French Straits	27.9	28.1	28.0	26.1

* Some of the passenger figures are calculated on the basis of 2.8 passengers per car equivalent.
† Vehicles are given different weights according to the space they occupy.
Source: *The Channel Tunnel: A United Kingdom Transport Cost Benefit Study* (A Report by Coopers and Lybrand Associates Ltd) London, HMSO (1973) tables 2.1 and 2.3

Only the principal producers and operators have been shown above. As will be indicated in chapter 14, however, there have been a large number of companies who have been involved with hovercraft for short periods. There are also a number of companies on the fringe of the industry, eg hovercraft consultancies, which have been set up by people who have at some stage been concerned with hovercraft.

11.2 Markets

The hovercraft concept as embodied in the first experimental craft, the SRN1, and its BHC descendants, has several technical features that distinguish it from conventional transport. Firstly, because the hull is completely separated from the surface by a cushion of air, this type of hovercraft is amphibious. This in turn means easier docking facilities and greater versatility, particularly over terrain that could not be reached either by boat or motor vehicle. At the same time, however, it creates special problems of manoeuverability (particularly at low speeds) and propulsion. Secondly, it is much faster than conventional marine craft. For example, the SRN6 can travel at 50–60 knots, while the conventional ferry travels at about 12–15 knots. The speed factor is also the principal advantage of the rigid sidewall craft now being produced by Hovermarine Transport and of the hybrid craft originally produced by Vosper Thornycroft which had no rigid sidewalls but was propelled by marine screws. Higher speed does of course mean higher utilisation rates. As the sidewall craft are not amphibious,* their speed is somewhat slower than their amphibious counterparts. Vosper Thornycroft's craft was only semi-amphibious because of the skegs to which the marine screws were attached. Its speed was also slower.

In the early days of hovercraft the civil market, particularly for commercial ferries, was seen as the most important. The defence departments had decided not to continue with hovercraft development (see chapter 12 below). The NRDC then took over sponsorship of the work, and in its report for 1958–59 stated:

> The Corporation is continuing to support the [hovercraft] project in the belief that the novel means of transportation represented by 'Hovercraft' has a significant part to play in the future in cargo and passenger carrying on water and overground.[2]

In 1961 a director of HDL, outlining NRDC policy, stated that the aim was:

> ... to encourage industry to produce hovercraft capable of being used *in the first place for commercial transport* and which could stimulate an overseas demand while enabling the Service departments to study ... how the new form of transport might be applied to their purposes[3] [author's italics].

Annual Reports of NRDC in the early 1960s stressed the need of the project to meet commercial requirements, and consideration was given to subsidising operators who were willing to use hovercraft in commercial operations. In 1965, the (now) managing director of BHC stressed the *civil* objectives of the early programme:

> The main hovercraft development effort has so far been directed mainly towards the civil market. Although the fighting services

---

* When this craft is riding on its air cushion the sidewalls still remain in the water and, with the flexible skirts at both ends of the craft, act to retain the cushion pressure.

have made use of machines like the SRN3 and SRN5, these have been specially adapted for their purpose and were not designed at the outset as their machines.[4]

This emphasis in the early days of the British industry on *commercial* applications of hovercraft in the civil market was in marked contrast to the American policy, which was (and still is) to concentrate its attention on military uses. It was perhaps inevitable, given the attitude of the defence authorities: the NRDC could hardly support the development of a military craft which the relevant government departments had already rejected. The bias of the British industry was shown in 1963, when Westland (which later merged with Vickers to form BHC) announced firm prices and delivery times for the SRN2, 3 and 5 in the hopes of attracting commercial orders. No SRN2s or 3s were sold, however, to commercial operators.

It is interesting to note in the light of these comments that the only commercial hovercraft routes in this country that have survived for any length of time are across the Solent and the Channel. These involve five SRN4s (the total number built) but only four of the fifty-three SRN5s and SRN6s built to date. By the end of 1973 only ten other SRN6s had been employed for civil purposes at some stage of their lives. It is clear therefore that for BHC the emphasis has shifted towards non-commercial applications of amphibious craft. Its latest craft type, the BH7, was designed as a military craft from the outset.

Vosper Thornycroft also aimed in the first instance at commercial applications for its hovercraft. It has since decided to concentrate its sales efforts on military markets. For this purpose it is developing a second generation craft, VT2 (as a private venture). The advantages of hovercraft even in military roles over their nearest competitor, the fast patrol boat, are not however seen by Vosper Thornycroft as overwhelming. In several aspects they come out worse in comparison. It is the specific features of air cushion vehicles, which are not available with conventional marine craft, that give them any advantage they may have.[5]

It is difficult to be conclusive over why these changes have come about, but as far as existing commercial routes that are not operating hovercraft are concerned, it appears that neither BHC's nor Vosper Thornycroft's craft are currently competitive or suitable. In 1970 the Programmes Analysis Unit published a summary of a study that it had undertaken into the prospects for high-speed marine craft (hovercraft and hydrofoils). The Unit argued that:

> ... The foreseeable costs of operating these craft indicate that, in general, it will be necessary to charge premium fares compared with competition from other slower transporation systems. However beyond distances of about 100 miles it is probable that competition from the air will be superior at such fare levels.[6]

Even on the high-density Southampton–Cowes route, BR Hovercraft is still losing heavily after eight years of operation (but see p. 294). (It is of course compatible with this overall view of the lack of competitiveness of currently available craft that *some* routes may be commercially viable.)

It may be significant when analysing the change in orientation that Westland, its predecessor Saunders-Roe, and Vosper Thornycroft are all companies that have been largely geared to military needs and markets throughout their existence.[7] This background may have led these companies in developing their hovercraft to concentrate rather more on technical excellence than on the realities of commercial ferry operation.[8] Vosper, for example, has a long history of building high-speed military patrol boats. And Westland's main product, the helicopter, is much more extensively used in military and para-military situations than in commercial operations. Saunders-Roe's work had been largely of an experimental nature, financed by government contracts and usually related to military needs. It is of course too early to be conclusive on this matter, but it would indeed be strange if a company's background did not strongly influence the way in which it tackled new projects.

The switch to military from commercial markets also raises questions about NRDC's involvement. Should it have entered hovercraft development at all? If so, did it back the wrong research projects? The latter question raises the further issues of whether *better* decisions could have been made on the *then available information*. In this respect, it is interesting to note the comments (recently made to the author) by one person closely involved with HDL's technical work: 'Even with several additional years of hindsight, it is still not obvious which of the R & D items should have been concentrated upon in order to maximise the return' (see chapter 15). However, without access to HDL's own evaluations of hovercraft prospects during its life, it is not possible to assess how valid this view is.

242 CASE STUDY : DEVELOPMENT OF THE HOVERCRAFT

The other major contender for hovercraft markets at present (1973) is Hovermarine Transport. Its product is clearly directed towards the civil market, but its marketing strategy is firmly based on the principle that its sidewall craft represent not a fundamental change from existing forms of craft but rather an improvement on the latter. The chairman of the Company stated at the end of 1970 :

> Our challenge is to convince our markets . . . that our craft is a relatively minor operational departure from the slower vessels that potential customers currently operate and maintain. That is, sidewall hovercraft form an evolutionary, not a revolutionary, step in marine transportation.[9]

The craft's sales are now increasing, although it is still too early to say how successful it will be. The craft of Air Bearings, Air Vehicles, Sealand Hovercraft and Enfield Marine are also aimed principally at the civil (but not necessarily the commercial) market, although they may have military applications. The first three, however, are much smaller than anything BHC has produced, and Enfield Marine's purpose-built freight craft is currently in a class of its own.

Export markets for hovercraft now appear to be the main area for expansion. Although all the SRN4s produced so far have been bought for use in this country, thirty-three of the SRN5s and 6s have been sold abroad. Nearly all Hovermarine Transport's output to date has gone abroad. It is probably true to say that most companies now see any future expansion as being overseas.

NOTES on Chapter 11

1. Chairman of Westland, quoted in *Hoverfoil News* (31 January 1974)
2. NRDC *Annual Report for 1958–59* (HC 38, Session 1959/60) London, HMSO (1959)
3. L. A. Sweny, 'Progress with Hovercraft' *New Scientist* (February 1961)
4. R. Stanton Jones 'Towards a Big Hovercraft' *New Scientist* (February 1965)
5. Report of a lecture by Mr P. J. Usher to the United Kingdom Hovercraft Society, reported in *Hoverfoil News* (30 December 1972). In 1969 the Navy's view was that, although they had 'a close interest in the very high speeds over water of which hovercraft are capable', hovercraft 'at present . . . have only a small payload in relation to the size and cost and their range and performance in moderate sea conditions . . . are limiting factors in maritime operations'. Memorandum

by the Ministry of Defence *Second Report* Select Committee on Science and Technology (session 1968/69) (HC 213) London, HMSO (1969) p. 213
6. K. M. Hill 'A Note on Some Economic Considerations for Government Policy towards Hovercraft and Hydrofoil' *Hoveringcraft and Hydrofoil* (December 1970)
7. For a history of Thornycroft before its merger with Vosper see K. C. Barnaby *100 Years of Specialised Shipbuilding and Engineering* London, Hutchinson (1964). For a history of Vosper see C. Dawson *A Quest for Speed at Sea* London, Hutchinson (1972). While both accounts were written specifically to celebrate the companies' centenaries, they nevertheless give a good guide to their orientation.
8. See C. Dawson op. cit.
9. Hovermarine Transport Ltd *Annual Report 1970*

12 The Individual Inventor

Cockerell was a trained electronics engineer who had worked for Marconi's for sixteen years before deciding to move into caravan sales in 1951, aided by a legacy left to his wife. During his time at Marconi's he had been responsible for several patents. Attached to his newly acquired caravan interest was a small boat-hiring business, and when selling caravans became unprofitable he decided to concentrate on boat-building. Once this business was established, he started experiments aimed at reducing the resistance that craft encountered when moving through water. He examined the effects of pumping a film of air under the boat on a theoretical and empirical basis, and began to experiment with various methods of containing the cushion. He started with fixed sidewalls with hinged doors at the ends, with air being pumped into the chamber under the hull. Towards the end of 1954 he tried replacing the hinged doors with cushions of water. He eventually hit on the idea of using a direct air jet instead of the sidewalls and water curtains, and tested the possibilities out by using a simple home-made rig.[1] Several further experiments followed, and a model was constructed with the help of a fellow boat-builder which incorporated the circular jet. Cockerell filed his original provisional patent application in December 1955 on the basis of these experiments. The abridgement of the final patent specification describes a vehicle.

> . . . which comprises means for causing a fluid to issue from the lower part of the vehicle in such a way as to result in the formation and maintenance of at least one curtain of moving fluid which travels across the gap that in operation exists between the surface over which the vehicle is to hover or travel and the structure of the vehicle and together with the said structure or surface encloses a space into which the said fluid or a gas other than the said fluid flows so as to result in the formation of a pressurised cushion or cushions by which the vehicle is wholly or partly supported . . . the fluid forming the curtain may be

liquid or gaseous and in one example an air jet is projected through a substantially circular orifice on the vehicle undersurface.²

Several comments may be made on Cockerell's invention and the events leading up to it. Firstly, his original work was based on a painstaking 'little-by-little' approach to a problem which he thought needed solving. His own comments on the period bear this out: 'It happened by lots of simple steps. It was just a case of plodding along trying to solve a problem which one had set oneself'.³ Numerous acts of insight varying in their technical importance occurred. In this sense Cockerell's work supports Usher's analysis of invention (see p. 33). Secondly, although Cockerell's work was undertaken in the context of his own boat-building business, it seems that his principal motivation lay in seeking a solution to a long-standing technical problem rather than in achieving economic success. Thirdly, Cockerell was a qualified engineer who had already been engaged in inventive activity (although not in the same field), he was not one of Schmookler's 'untrained' inventors (p. 63) Fourthly, in comparison with subsequent development, the costs of his original experiments up to the time of his first provisional patent applications were very low – probably only a few hundred pounds at the most. This compares with £8,000 spent by the (then) Ministry of Supply in evaluating the basic idea and the £120,000 spent by HDL on building and developing the SRN1. Even these sums are small compared with later expenditure: BHC has probably spent well over £15m. (from various sources) on R & D and on the construction and testing of prototype craft.

Jewkes *et al.* have of course used Cockerell's work as an example of the independent inventor working on his own meagre resources who has produced an important technical development.⁴ This view is undoubtedly correct and should reinforce support for the place of the indepndent inventor as a source of technical advance even in the present technological environment. However it should be remembered that, without the following R & D programmes for which Cockerell's work was the initial stimulus, the general principles utilised by him would never have been developed. This is particularly important to remember especially as the original annular jet proposed by him has since been largely superseded by subsequent technical developments. Both elements were equally necessary and it is a futile exercise to apportion all the credit to

246 CASE STUDY : DEVELOPMENT OF THE HOVERCRAFT

one or the other.

Finally, Cockerell's work was not dependent on identifiable preceding scientific work. Indeed, nearly all of the more basic research on the nature of the air cushion was undertaken *after* Cockerell had demonstrated the practicality of his ideas. This is another example of the way in which the basic assumption underlying the Byatt–Cohen method of basic research evaluation (see p. 217) may not always hold.

After his provisional patent application Cockerell continued his work on the air cushion principle and sought ways to refine it (he filed his complete application in December 1956). During this time he gave some consideration to the use of flexible extensions to the air cushion. In May 1957 he filed three more provisional patent specifications based on this further work.

After Cockerell had built the home-made rig, (in the summer of 1955), he sought support for his idea in industry. He approached several firms but without success. English Electric turned it down because they could not envisage an aircraft application for the idea. Early in 1956 Napiers also decided that they could not back the project, because it was at such an early stage of development. Later that year De Havilland's turned the proposal down. Eventually, through his landlord, Lord Somerleyton, Cockerell was able to enlist the support of Lord Mountbatten, who asked the Admiralty to examine his idea more closely.[5] In December 1956 the Admiralty invited Cockerell to demonstrate and explain his ideas. At that meeting was an assistant director of research at the Ministry of Supply, Ronald Shaw, who had already had experience with flying boats and had in fact undertaken some limited experiments on air lubrication himself.[6] No one in the Admiralty was sufficiently enthusiastic to take the matter further, but Shaw invited Cockerell to further discussions. As a result of these discussions Cockerell became a consultant to the Ministry and the latter eventually (in October 1957), placed a small evaluation contract with Saunders-Roe on the Isle of Wight (again, only after Short Brothers had turned the contract down on the grounds that they could not see a worthwhile application in their field of business). Saunders-Roe was asked to evaluate the principle and to indicate where hovercraft might fit in to the overall transport context. The Ministry also classified hovercraft work because of its possible defence applications.

Nearly a year later, while Saunders-Roe were still working on their contract, Cockerell was faced with the need to file complete specifications in overseas countries if the priority date of May 1957 for his patent applications was to hold. This priority date was particularly important because a Swiss inventor, Carl Weiland, was working at the same time on similar lines and had filed a provisional patent specification in October 1957. Once again, Cockerell turned to English Electric to see if he could gain financial backing to secure patent protection in Europe and to develop his ideas further. Lord Caldecote was asked by Sir George Nelson, the Chairman of English Electric, to see Cockerell and to make a recommendation on whether English Electric should become involved in the project. As a result of this meeting, Caldecote concluded that 'the project was at too early a stage to make it suitable to be taken up by English Electric since it would have been an entirely new kind of product for the company which was already fully stretched in development within its existing field'.[7] He did however, suggest that Cockerell should approach the NRDC, which the former did in April 1958. The NRDC decided to provide the necessary finance for the patent filings in exchange for the first option on the invention. In May 1958 the Ministry of Supply received the report from Saunders-Roe. The conclusions of the report were broadly favourable, although lacking in any quantitative evaluation of possible markets. The Ministry decided not to pursue the project in any substantial form despite Shaw's own enthusiasm, and it was declassified. The Ministry's decision must be seen in the light of the changes brought about by the 1957 Defence White Paper (see chapter 15). The aircraft interests in the Ministry were dominant and the contractions outlined in the White Paper made finance for new projects outside aircraft difficult to obtain.

After completion of the Saunders-Roe contract, a proposal was put up by those who worked on the contract to the company's main board, proposing a privately financed project for building a manned craft. This was turned down by the board for financial reasons. NRDC then decided, in September 1958, primarily on the basis of the evaluation report, to back the project by placing a contract with Saunders-Roe to design and build a man-carrying experimental craft. The proposal on how this should be done was put forward by Saunders-Roe. This proposal differed very little from that which had been put before the Saunders-Roe board in the previous months. In January 1959 NRDC set up a subsidiary, Hovercraft Development Ltd, to monitor the contract's progress. Both Cockerell and Shaw were made board members of the new company.

It is tempting to use Cockerell's experience of trying to obtain acceptance for his invention as an example of the type of situation for which the NRDC is ideally suited; that NRDC's action was, in the words of a government minister in 1966, 'an outstanding example of the benefits of public intervention in this field'.[8] As we have seen however, subsequent experience has so far left it an open question whether the NRDC's involvement was economically justified at that stage (see also chapter 15). It must also be remembered that the existence of the Saunders-Roe report to the Ministry of Supply inevitably affected NRDC's attitude towards the air cushion. What would have happened in the absence of a report from a technically reputable company is of course speculation, but NRDC may well have decided against the project, especially in view of its previous experience of proposals put up by individual inventors, few of which seemed to have had potential (see p. 145)

Two other observations are worth making on Cockerell's early experience. Firstly, he did not have the personal resources to take advantage of international patent protection. The cost would have been in excess of £1000.[9] Secondly, from conversations with people involved at the time, and from published accounts of what happened, a strong impression is gained of the importance of personalities and of the crucial consequences of 'the right people being in the right place at the right time'. Apart from Cockerell's own enthusiasm and persistence in the face of industry's lack of enthusiasm, the interest and involvement of others was also significant. Shaw, for example, was one of the few to show enthusiasm for Cockerell's work in the early days – but for his presence at the initial meeting with Cockerell the original evaluation contract might not have been placed.

Cockerell's inventive ability and commitment to hovercraft continued to show itself at HDL. Of the 154 patents taken out by or assigned to HDL between 1957 and 1967, over fifty have Cockerell's name on them as sole or joint inventor (see however pp. 302–3). In 1969 a biographical note commented:

> ... he was looked upon by his staff ... as being very approachable and could be relied upon to lend a sympathetic ear to any ideas which were put to him. He did perhaps at times transmit the popular image of the inventor, that of coming up with ideas

which were essentially intuitively based, though frequently sound, and leaving his staff to work out the theory as to why it worked and how it could best be applied. In this way, his enthusiasm and capacity for thinking up these ideas tended at times to overload the engineers at HDL.[10]

In the early 1960s, under Cockerell's direction, the work of the Technical Group at HDL embraced a wide range of applications of the air cushion principle. For example, sidewall and tracked hovercraft and industrial applications were all areas to which resources were devoted. By 1966 however enthusiasm for hovercraft development was beginning to wane. In that year, with NRDC's blessing, Vickers and Westland, the two main companies involved in hovercraft work, merged (see chapter 14), thereby forming a virtual monopoly of hovercraft activity in this country. Cockerell strongly opposed this move, arguing that competition was essential for rapid technological progress in this field. Rightly or wrongly, however, Cockerell was not always regarded as having sound commercial judgement. In 1968 the then Minister of Technology stated : 'Mr. Cockerell is a very capable inventor . . . I think that some people in the NRDC could not regard him as quite so high in the business sense.'[11] Cockerell resigned from HDL in 1966 and his links with the established hovercraft industry gradually lessened. Until recently, he has been working on his own financing his studies largely out of his own resources. In 1973 however BHC appointed him as a consultant.

It is clear that Cockerell has always been and remains firmly convinced about the eventual commercial competitiveness of hovercraft and of its wide applicability. To him, shortcomings in the government framework for supporting major inventions are, in part at least, responsible for the present unenthusiastic attitude towards hovercraft rather than any lack of basic potential in the air cushion itself.[12] Whether or not he is right in this respect, it is very clear that without his initial vision, his complete commitment to his ideas and his refusal to be put off by the repeated displays of industry's lack of interest, the project would not have been taken up.

NOTES on Chapter 12

1. 'The Cockerell Papers' *Flight* (February 1963)
2. Abridgement of patent no. 854211
3. Quoted in *Flight* (February 1963)
4. Jewkes *et al. The Sources of Invention* 2nd edn. London, Macmillan (1968)
5. P. Fairley *British Inventions in the 20th Century* London, Hart Davis (1972) p. 87. This was confirmed in writing to the author by Lord Mountbatten.
6. B. Cooper 'The Hovercraft Pioneers' *Hoveringcraft and Hydrofoil* (September 1967) and interview with R. A. Shaw (July 1973)
7. Letter from Lord Caldecote to the author (8 July 1974)
8. *Hansard* (Commons) Col. 1098 (12 February 1966)
9. P. Fairley op. cit. p. 82
10. 'The Hovercraft Pioneers: Christopher Cockerell', *Hoveringcraft and Hydrofoil* (February 1969)
11. *Hansard* (Commons) Col. 902 (28 February 1966)
12. 'Launching Major Projects' (memorandum by Sir Christopher Cockerell to the Select Committee on Science and Technology) *Tracked Hovercraft Limited* (Appendices to the Minutes of Evidence, HC150 – ix Session 1972–73) London, HMSO (1973) pp. 11–13

13 The Technical Development of the Air Cushion Principle

This chapter surveys the way in which the initial concept of the air cushion has been extended and revised. The first section gives a brief history of technical development in the industry. The second section analyses inventive activity by using patent data. In the third section the craft types so far produced are discussed. The final section looks briefly at technological spin-off from the air cushion principle.

13.1 Technical history

13.1.1 *Early work on the air cushion*

Although it is clear that British technical development of hovercraft started from the evaluation by Saunders-Roe of Cockerell's early work in 1958, there had been numerous earlier attempts in various countries since the mid-nineteenth century to produce a lubricant of air between the hull of a boat and the surface of the water.[1] Some of these efforts led to patents. For example, John Thornycroft took out a patent in 1877 which covered the creation of a cushion of air between the bottom of the boat and the water surface. He built models for his experiments but went no further. (Thornycroft was the founder of the Thornycroft shipbuilding firm that was taken over by Vosper in 1966. By coincidence, Vosper Thornycroft is now involved in hovercraft activity (see section 13.4).) A few years later a Swede, Gustav de Laval, obtained a patent for a ship that incorporated an air-ducting system in its hull: air bubbles were forced through tubes under the water line, the intention being to provide an air cushion on which the hull would ride. Tests proved unsuccessful however and the project was discontinued. Experiments on the use of the air cushion in different forms and in various types of craft continued intermittently up to the 1940s. For example, an experimental plenum chamber craft was built in Finland in the 1930s.[2] (In a plenum chamber craft, air is pumped into a cavity on the underside of the hull to form a cushion of high pressure. The air then leaks out under the edges.) None of these got much beyond the prototype stage, however.

In the 1950s, when Cockerell was working on his experiments, there were others who were thinking on similar lines. For example, a Brazilian filed a provisional specification covering a type of air cushion vehicle in 1955. The vehicle relied for its cushion on a peripheral jet 'to create "a column of air which reacted against confinement" and on which the vehicle rose 15–20cms above the ground'.[3] However, because of lack of interest the inventor did not pursue his application. In the mid–1950s also Carl Weiland was trying out his ideas in Switzerland. His particular invention related to 'labyrinth seals' which involved a complex arrangement of circular channels on the craft's underside through which the air is forced. The air is re-energised in each channel.[4]

In the United States work was also being done on the air cushion. At a relatively unimportant level, Hoover had already patented its air cushion-supported vacuum cleaner in 1954. By 1957 Ford had started work on its 'levapad' system, which involved the provision of a thin film of air between the surface (which had to be very smooth) and several 'pads' attached to a platform. In the mid-fifties other work on the air cushion and the circular jet was also being carried on in the US.[5] For example, in April 1957 Colonel Melvin Beardsley, who had been working on the circular jet, filed a patent similar to that of Cockerell's original application. However the contest was later fought out and won by the NRDC. In 1959 the Curtiss-Wright Corporation tested its 'air car' prototype which was designed to operate approximately one foot from land or water surfaces and designed for speeds of up to 60 mph. The interesting feature about this vehicle in view of later technical developments was that it incorporated a flexible skirt 'which acts as a boundary for air flow' and 'permits the vehicle to pass over low obstacles'.[6]

The above survey is not intended to be exhaustive, but the examples do make it clear that Cockerell's own investigations had been preceded by experiments conducted elsewhere along similar lines, and that others unconnected with Cockerell were working in the field at the same time as he was. The possibility of using an air cushion as a means of reducing resistance was not therefore an entirely new idea, although the amount of effort that had gone into exploring the concept further was minute compared with the R & D programme of the last fifteen years.

Why was it that Cockerell's work was eventually taken up while that of his predecessors eventually foundered? Cockerell's major contribution of the peripheral jets gave his ideas a clear technical distinctiveness in comparison with earlier work. Other factors, which individually may have seemed incidental, also combined to give initial impetus to the work. It is worth noting, for example, that at about the time that Cockerell's proposals were being examined, the vertical take-off aircraft was being developed. With this aircraft, an *adverse* ground effect was experienced with take-off, requiring power for lift many times greater than for forward propulsion. Cockerell, however, was in essence proposing that the ground effect could actually *assist* lift.

It may of course be argued, more fundamentally, that technical conditions were ready for this development, particularly as limits on the speeds obtainable by conventional boats were being reached. The parallel experiments in different countries in the 1950s give some, though weak, support to this hypothesis. The acceptance of Cockerell's work must also be seen in the light of his own persistence and enthusiasm discussed in the previous chapter : while this was clearly not a *sufficient* condition for subsequent exploitation, it was certainly a *necessary* element.

13.1.2 *Subsequent development*

Cockerell's original system of peripheral air jets has to some extent been superseded by subsequent technical developments in the industry. One of the problems that Cockerell and the early investigators faced arose because the clearance between the craft and the ground was only very small. This meant that the hovercraft could be used only on fairly flat surfaces and not, for example, in seas where the waves were above a certain height – the SRN1 Mark 1 could not operate in seas with wave heights above eighteen inches. Cockerell had himself given some thought to this problem and in

1958 had talked about some kind of flexible arrangement 'to enable the bottom of the craft to conform to the surface beneath it'.[7] He did not however get to the stage of patenting his ideas. The French company, Bertin et Cie, had also examined the possibilities of surrounding air jets with 'petticoats'. However, the SRN1 was originally designed and built without skirts.

Much of the technical effort since 1959 by HDL and the companies involved has been devoted to solving the problem of moving over rough surfaces by means of flexible skirts attached round the periphery of the craft. Indeed, the technical history of the hovercraft centres largely on skirt development. At first (1960) flexible extensions were attached to the peripheral jets on the SRN1. However because of the problems of maintaining the angle of the jet with these extensions, despite the use of chains and diaphragms, the peripheral air jet concept in its original form was abandoned in favour of other systems (see below).

Since the days when the first jetted extensions were fitted, HDL and BHC* have followed somewhat different lines of development in their skirt systems. BHC first developed a peripheral bag with a jet nozzle on its underside. This arrangement proved unsatisfactory from a wear and repair viewpoint and in 1966 fingers replaced the nozzle. These fingers were separate members, and in the event of wear or damage could be replaced individually. Today, on most BHC craft the finger accounts for about fifty per cent of the cushion depth, the bag providing the rest. HDL on the other hand developed a fingered (or segmented) skirt which was attached to the craft structure by means of a loop. No bag is necessary in this system. The BHC skirt requires thicker material than the HDL skirt. Vickers also produced its own convoluted skirt which is very similar to the HDL design: the latter is regarded by some as a direct descendant of Vickers' work.

In the early 1960s, BHC and Vickers were each proclaiming the merits of their respective skirt systems. In a press announcement in 1963, Westland referred to its skirt development as 'an important breakthrough' and went on: 'Westland ACVs will now have a considerable amphibious performance without an appropriate increase in power or size. This outstanding advantage can be applied to smaller craft.'[8] Four months later, the manager of Vickers'

* The term BHC here covers Westland before 1966

Hovercraft Division referred in an article to 'the inflated type of skirt' of the VA2, which 'incorporates a technical advance *over all other skirts*'[9] (author's italics). Competition between the two companies certainly appeared to foster separate lines of technical development.

It would be presumptuous for the author to provide a detailed technical evaluation of the two main skirt systems. However both Hovermarine Transport and Vosper are licensed under HDL rather than BHC and utilise (in modified form) the former's skirt systems. This may however partly reflect the fact that both companies employed ex-HDL personnel as their key technical men in the initial phase of their hovercraft work. Enfield Marine, Air Bearings, Air Vehicles and Sealand Hovercraft also employ the basic HDL system, as did Cushioncraft.

The key patents in skirt development are difficult to pinpoint and are subject to dispute in the industry. Two however were invariably mentioned by people who discussed the matter with the author. The provisional specification for the first was filed in June 1958 by C. H. Latimer-Needham, an independent consulting engineer whose interest had been fired by reading accounts of Cockerell's work.[10] HDL opposed the granting of this patent, arguing that its claims were anticipated in Cockerell's orinigal specification. The patent was eventually granted in 1961 having been bought by BHC. Although this company had done some experimental work on skirts with the SRN1, under its (then) chief designer, R. Stanton Jones, they had no patent coverage for it. The purchase of this patent considerably strengthened their technical position. The second major skirt patent was applied for in 1961 by HDL, D. Bliss being named on the application. This specification and subsequent additions to it are regarded as crucial in the protection of HDL's skirt system. Thus the three major inventions in hovercraft, those of Cockerell, Latimer-Needham and Bliss, all come under the Jewkes definition of the 'individual' invention (see pp. 59–60).

It is misleading however to concentrate on key patents in skirt development. Throughout the 1960s, continuous improvements have been made to skirt life, through better materials, design etc, as shown in table 13.1. Skirt development has in turn generated its own problems, for example the tendency of early craft to 'plough in' and 'bounce'. Modifications in skirt designs have been necessary to

256 CASE STUDY : DEVELOPMENT OF THE HOVERCRAFT

Table 13.1
Replacement lives of skirt components
(hours)

	1964	1968	1973
SRN6 operating at 45–30 knots in calm to 4ft seas			
Main bags	500	2000	7000
Fingers	50	300	700
SRN4 operating at 65–45 knots in calm to 8ft seas			
Main bags		300	2000
Fingers		50	300

Source: BHC

overcome these problems. Parallel with skirt development has been the improved knowledge of the behaviour of the cushion under different conditions.

Hovercraft development also raised numerous questions on which relatively established technologies had some bearing. For example : How were the craft to be propelled? How was directional control to be obtained? What materials should be employed? What engines should be used? These questions had fairly obvious answers in some cases. For example, with sidewall hovercraft, both propulsion and steering could be of the conventional marine form. With amphibious hovercraft however there were a number of possibilities for propulsion. In the case of BHC it is likely that its own aircraft background had a strong influence on its hovercraft design philosophy; all its craft, apart from the SRN1, have been propelled by air screws. Vosper's first hovercraft type was propelled by marine screws, a propulsion with which they were of course familiar. Other forms of propulsion, using fans, have been employed mainly in small craft – the small company Cushioncraft (see chapter 14) used this method on its craft. Fans are also used for propulsion on craft produced by Air Vehicles and Sealand Hovercraft. This form of propulsion makes less noise than air screws.

Directional control on amphibious hovercraft presents a number of special problems, particularly in relation to shortcomings in the use of mounted fins and rudders, and various innovations have been introduced in an attempt to overcome them. Propellers mounted on swivelling pylons for example were introduced by BHC on the SRN2 in 1960. Later developments have included the provision of control ports on the sides of the craft and of devices that modify cushion pressures, for example by local lifting of the skirt.[11]

These later developments were particularly helpful in the case of craft propelled by single, fixed propellers.

As far as materials are concerned, most craft have been built with alloys. The D2 and HM2 however, together with the later smaller craft SH2 and AV2,* have been built with glass-reinforced plastics. BHC has until recently used only alloys, again materials with which it was most familiar. In many instances, new materials and components have had to be made to suit the special conditions of hovercraft operation. Salt ingestion by the engines (which is related to the craft's cushion pressure and the height of the air intakes above the water surface) and corrosion on propellers caused numerous problems in the early days. In consequence, new filters and protective coatings respectively have been developed. The engines used in hovercraft have varied widely. They have however all been adaptations of engines already used in aircraft, ships or cars. Engines specific to hovercraft have not been produced.

It is very difficult to disentangle the contribution of the different hovercraft manufacturers, operators and component suppliers in these fields, especially as technical progress seems to have consisted largely of steady and continuous improvement rather than of discrete technical jumps. There can be little doubt that both BHC's and HDL's contributions have been considerable. Unfortunately however we do not have the data to compare the *productivity* of the technical efforts of the different organisations involved. Furthermore, as far as BHC's work is concerned, it must be remembered that it has not enabled the company to achieve a substantial market in commercial ferries, its first objective.

One thing however is clear from the industry's technical history : there is no obvious break between inventive activity and the process of innovation, as suggested by Schumpeter (p. 19). Instead, inventive work has been an integral part of the attempt to market the hovercraft. There has been a constant stream of technical improvements designed to iron out the 'bugs' discovered in the craft initially put on the market and to improve their performance. Thus, the technical history has been made up of a process of invention – innovation – invention – innovation, and so on. To this extent, it supports Ruttan's thesis (see p. 30).

---

* For the companies producing these craft see notes to table 13.2

13.2 Inventive activity

The growth of inventive activity since 1959 in different companies and organisations as measured by patents is shown in figure 13.1. Although there are patents for earlier years, 1959 is chosen as the effective starting date of British activity. The graph is based on patents included under the Patent Office classification, *B7K : Air Cushion Vehicles*. As such the patents analysed cover a wide field – for example, many patents relating specifically to tracked hovercraft are included – and patents are frequently classified to other headings as well. Patents are allotted to the year in which the technical activity giving rise to the patent occurred.

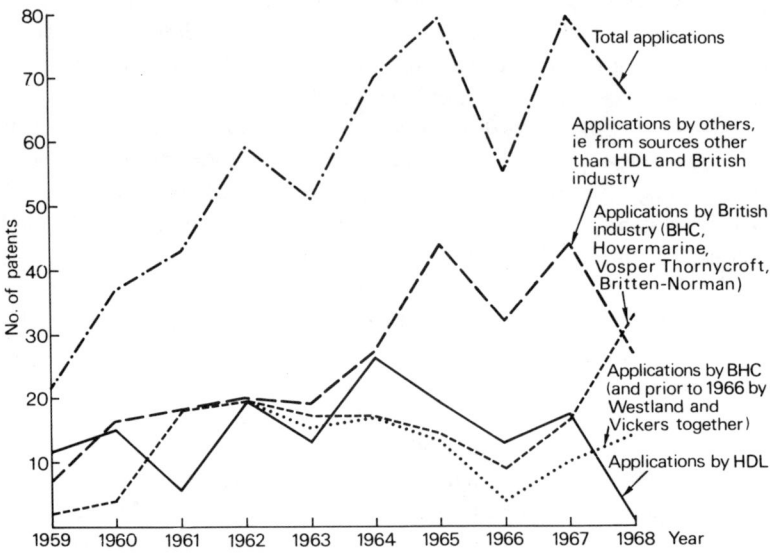

FIGURE 13.1. Provisional applications under Patent Office classification B7K : *Air Cushion Vehicles*.

Extensive criticism has been made of the use of patent data as a measure of technical activity (see pp. 33–5). This must be remembered when examining the data. As far as propensity to patent is concerned, however, it is clear from discussions that the author has had that both HDL and BHC were very patent-conscious. (BHC for example runs its own patents department.) This in turn has had its effect on other firms in the industry.

The data used in the graph were obtained from completed patent specifications; hence the eventual assignee named in this specification may not be the original inventor. However, this is unlikely to be an important source of error in the allocation of patents to different organisations. Because the year of the provisional specification filing is used, 1968 is unfortunately the latest year for which complete data are available. Some 1969 filings have not yet (January 1974) been accepted.

The output of inventive activity as measured by patents will be determined by two principal factors: firstly, the 'inputs' for that activity, namely R & D, and secondly, the 'productivity' of those inputs in terms of patents. Both will be influenced by numerous factors. Patents per unit of input will depend *inter alia* on technological opportunity in the fields investigated and on the extent to which results from different types of work can be put into patentable form. A fall in patents with no fall in input *may* merely reflect a shift in the type of work undertaken.

HDL's inventive activity rose rather jerkily to a peak in 1964, and then declined in 1965 and 1966 (1967 of course marked the end of its Technical Group). It is difficult to find an explanation for the yearly variation in HDL's inventive work from variations in its input: the staff grew linearly up to 1964 and then remained constant (at around the 100 level) until 1967. One possible explanation is that the frequent changes in its management and organisation may have had an influence on the amount of effort applied to patenting activities. It is interesting to note however that the inventive activity of BHC, and prior to 1966 of Westland and Vickers together, had peaks and troughs in the same year as HDL's (though less marked), suggesting perhaps that common influences were at work. Unfortunately as no commercial sales occurred before 1965, sales data cannot be used as a proxy for economic prospects to apply the tests used by Schmookler (see p. 37). As patent data for years after 1968 accumulate such tests should become possible, although they will still not give any direct insight into the economic influences affecting patent activity in the *early* years of a technological development, before any sales have been achieved.

The decline in HDL patent applications after 1967, which was not matched by any increase in NPL applications, is particularly interesting, especially as applications from British industry *rose* at the same time. Thus the take-over by NPL may have reduced inventive activity at a time when British industry was itself becoming more enthusiastic.

BHC's inventive activity declines after 1962, slightly up to 1964, then more rapidly to 1966, despite rapid growth of patents elsewhere between 1963 and 1965. Since 1966 it has again started to rise. No simple explanation exists for this behaviour. However one possibility is that the data reflect in part the changing structure of competition within the industry. Up to 1962 four firms, two of which are included under the BHC heading, were competing actively in the hovercraft industry. By 1963 however two had dropped out and the third was beginning to lose interest (it launched no craft after 1963). By 1966, when BHC was formed, Westland was the clear industrial leader (see chapter 14). After 1966 however new entrants to the industry appeared.

On the basis of these patent data, the relationship between the inventive activity, as measured by patents, of HDL and similar activity in British industry was explored using linear regression analysis. It will be remembered that one of the original reasons for setting up HDL was to encourage hovercraft interest and activity in industry. It seems reasonable therefore to examine whether HDL's work stimulated industry's own activity. It might be argued conversely that increases/decreases in HDL's work would lead to *counteracting* change in industry, with firms taking the attitude that they do not have to undertake the work that another body is already doing (see p. 132). However this is a weaker argument, at least in the field of patentable activity, since there is no guarantee that the results of such work would be generally available.

HDL's patent applications were lagged by one and two years but the results showed no evidence of HDL acting as a stimulus : with each lag, the r^2 is below 0.2, and the coefficient is not significant. Nor does there seem to be any support for the alternative possibility: that HDL's work *responded* to industrial activity.

In interpreting these results the limitations of the data in both quality and quantity terms must be borne in mind. The use of other data on technical activity, eg R & D expenditure, might yield different results; unfortunately no time series data on industrial R & D expenditure are obtainable for obvious reasons. Finance from public sources other than HDL, ie government departments, is also likely to have had some effect on industrial R & D activity; again

13.2.1 Craft types

Table 13.2 gives the main craft types produced over the period. Very small 'fun' hovercraft, the Folland ground effect research machine (GERM) and the Vickers Air Cushion Landrover are excluded. Craft are allocated to the year in which they are first launched. It should be remembered that some craft did not differ extensively from earlier types (eg the SRN3 was a military counterpart of the civil SRN2, and the SRN6 is a very close descendant of the SRN5). Craft may also be developed extensively after their initial launch. For example, the HM2 is now in the Mark III stage. The SRN1 went through five Marks with its weight rising from 3.8 to 6.7 tons. Several types, eg SRN6 and SRN4 have been 'stretched' to increase their carrying capacity. There is now a twin-propeller version of the SRN6.

On average, about two new craft types a year have been

Table 13.2
Craft launched 1959–72†

Year	Craft
1959	*SRN1 (1)
1960	CC1 (1); VA1 (1)
1961	CC2 (3); D1 (1); HD1 (1); VA2 (1)
1962	D2 (3); SRN2 (1); VA3 (1)
1963	SRN3 (1)
1964	CC4 (1); SRN5
1965	Hovercat (5); SRN6 (and SRN5 (53))
1966	CC5 (1); HD2 (1)
1967	CC7 (5)
1968	HM2 (18); SRN4 (5)
1969	BH7 (3); VT1 (3)
1970	AV2 (1)
1971	SH2 (7)
1972	EM2 (1)

* The prefix letters give the key to the producing company as follows:
  SRN and BH   Saunders-Roe (after 1959, Westland Ltd; after 1966, BHC)
  VA   Vickers   HM   Hovermarine
  CC   Cushioncraft   AV   Air Vehicle
  D   Denny Hovercraft   SH   Sealand Hovercraft
  HD   HDL   EM   Enfield Marine

The prototype Hovercat was launched by a small Isle of Man company. A version of this prototype was later produced by Hovermarine.
† Figures in brackets refer to the numbers produced.
Source: trade literature

produced. However very few of the twenty-five types named have been manufactured in any quantity. (Numbers produced are in brackets in the table.) None of the eleven craft types launched up to 1964 was developed for quantity production (some were intended only for research purposes). In nearly all cases only one craft was produced, although some of these, eg the SRN2 and VA2, were used on experimental passenger services. 1964, therefore seems to be the best candidate for the 'innovation' date, if such a date has to be identified; it was in this year that the first craft to be sold commercially was produced. Craft had of course been sold before that date. For example, the SRN3 was sold to the Ministry of Defence. Two CC2s had also been bought by the government. These craft however were prototypes, and were not put into quantity production.

If 1964 is taken as the innovation date, this gives a gap of nine years from the date of Cockerell's provisional specification. Comparative data against which to assess this lag between the invention and innovation of the hovercraft is difficult to obtain, but the limited evidence available (see pp. 173–4) suggests that nine years is by no means a long gestation period and is shorter than that for a wide range of other innovations. However, it must be remembered that this lag is determined by numerous factors and may not be a reflection solely of the commercial judgement of the firms involved. Government finance, for example, played an important part. Furthermore, as already mentioned, substantial technical development and 'trouble-shooting' has been necessary since 1964. Thus it would be misleading to attach too much importance to this year.

Most of the craft mentioned in table 13.2 have been assisted – by grant, purchase of prototype or loan – by the government or the NRDC (see chapter 15). Only the CC1 and 2, D2, SRN5 and 6, Hovercat, SH2 and EM2 have been entirely privately financed. It is especially noteworthy that the most successful craft to date, the SRN5 (then 6), was a *private* venture by BHC.

It is likely that new craft will continue to be produced in the near future. Vosper Thornycroft is already working on its VT2, a military version of the VT1, and Hovermarine Transport is undertaking preliminary design studies for its HM5. However, BHC currently has a policy of selling off the shelf and making 'steady improvements to the breed rather than trying to make massive steps into the unknown'.[12] This policy is in line with the cautious technical approach of BHC's parent company (see pp. 282–3). It may also reflect the drying up of government funds.

13.3 Technological spin-off

Early in its life, HDL had considered the possibilities of high-speed movement over land using the air cushion principle, and it took out several patents in this area. Others, particularly Ford in America, were also experimenting in this field.[13] One of the problems that HDL faced was over the choice of propulsion. It eventually decided to utilise the linear induction motor, and an experimental air cushion model was demonstrated in 1966. In 1967 NRDC allocated £2m. towards the development of tracked hovercraft, and because HDL's Technical Group was absorbed by the National Physical Laboratory in that year, an NRDC subsidiary, Tracked Hovercraft Ltd, was formed. The company was established at Cambridge and built its own track for experimental purposes. By 1971 however NRDC had become much less enthusiastic about the project and was coming to the view that the funding that THL required was beyond its own resources. In 1972 the government decided that no further money would be made available for THL, which was therefore wound up. This decision came in for very strong criticism from the Select Committee on Science and Technology pp. 149–50). Whether such criticism could be justified from the evidence before the Committe is however doubtful.

A second area of spin-off from hovercraft work has been in industrial applications. This has for instance involved the building of hover-platforms of different configurations for moving heavy loads. The air cushion has been applied in this way to move oil tanks, heavy generating equipment, damaged aircraft and dredgers. (One of the problems faced by offshore dredgers is that in only moderate seas they become very unstable. This usually means that they have to return to their base, which may be several miles away. Supported on an air cushion however they can come ashore at the

264 CASE STUDY : DEVELOPMENT OF THE HOVERCRAFT

nearest practical point to their operations and can recommence activities as soon as the seas are calmer.) Little new technology has been involved in these applications; rather existing technology has been applied to new situations. Other specialist areas in which the air cushion has been applied have been in the provision of 'beds' for the treatment of badly burned patients. This development, which is supported by NRDC, is still very much at the experimental stage. NRDC also gave some support to the development of a hover kiln, although this project was abandoned in March 1971.

To date therefore the spin-off from hovercraft work appears to be fairly limited. Apart from some industrial applications – and even here the market is currently very small – there has not been a very extensive application of the air cushion principle. The biggest offshoot, tracked hovercraft, is now virtually defunct.

NOTES on Chapter 13

1. The following brief survey relies heavily on L. Hayward *The History of Air Cushion Vehicles* London, Kalerghi-McLeavy (1963) and H. F. King *Aeromarine Origins* London, Putnam (1966)
2. *Air Cushion Vehicles: Transportation of the Future* New York, Transportation Research Associates (1962)
3. L. Hayward op. cit. p. 36
4. Egon Larson *Hovercraft and Hydrofoils* London, Dent (1970) pp. 42–3
5. T. Margerison 'Principles of the Hovercraft' *New Scientist* (11 June 1959)
6. E. J. Bulbin 'Curtiss Wright Tests Air Car Prototype' *Aviation Week* (1959)
7. *Ripplecraft Report* (March 1958) quoted in *Air Cushion Vehicles* (28 August 1963)
8. Quoted in *Air Cushion Vehicles* (23 May 1963)
9. S. R. Hughes 'Flexible Skirts for Air Cushion Vehicles' *Hoveringcraft and Hydrofoil* (September 1963)
10. C. H. Latimer Needham 'Early Days Recalled' *Hoverfoil News* (April 1973)
11. See G. H. Elsey and A. J. Devereux *Hovercraft Design and Construction* Newton Abbot, David and Charles (1968) chap. 8
12. Interavia *Air Letter* (1 March 1973)
13. D. Bliss 'The Evolution of Tracked Air Cushion Vehicles' *Hoveringcraft and Hydrofoils* (August 1970)

14 The Development of the Industry

This chapter examines the development of the industry. The first section analyses the growth of the manufacturing sector; the second section looks at the development of commercial hovercraft operations.

14.1 Manufacturers

14.1.1 *The beginnings of the industry*

The Ministry of Supply first offered the evaluation contract to Short Brothers at Belfast (see p. 246), a company that had had experience with flying boats. After Shorts had turned the contract down, Saunders-Roe was approached. Because of the cancellation of defence projects following the 1957 White Paper, the government had decided to stop the company's development of the SR177, a manned interceptor. The firm was therefore seeking work for its employees and for its specialist facilities, including its test tank. It was prepared 'to consider anything which showed a reasonable chance of commercial success and was within its capacity'.[1] The evaluation contract was therefore accepted, and it is from the Saunders-Roe base that the industry grew. In one sense, the company was the obvious choice for the later NRDC contract to build the SRN1; not only did it have previous experience of flying boats and of the earlier evaluation contract, but it had a long history of prototype building. Between 1945 and 1959 it had built the SRA1 experimental jet fighter, the Princess flying boat prototype, the SR53 aircraft and then the SR177 (a development of the

SR53). In another sense however, the choice was less obvious; as we have already pointed out (p. 241) it is possible that the company's prototype background in aircraft and its close link with military requirements may have led to a particular technological bias in the way in which hovercraft were initially developed which was not entirely appropriate for the commercial market. On the other hand, it must be remembered that at the moment the three principal commercial ferry routes in this country operate with craft produced by Saunders-Roe's descendent, BHC.

While the SRN1 was being built, the newly formed HDL was seeking licensees for the Cockerell patents. Its efforts were unsuccessful. Unfortunately, it has not been possible to find out with which companies discussions were held or the reasons why no agreements were signed. Cockerell's own experience however (p. 247) suggests that two factors are likely to have been particularly important in determining industry's attitude. Firstly, the project was still at a very early stage; and, secondly, it did not fit in easily to either aircraft or ship technology : very few firms straddled both. Because it was not immediately possible to find a firm or firms on to which the project could be grafted, HDL established its own Technical Group (eventually located at Hythe, Hampshire) in the hopes that potential licensees would be attracted by the technical services it offered.[2] Licensees were to be given the technical reports of HDL. This procedure was not the normal policy of NRDC in sponsoring an invention, but the corporation took the view that because of the unwillingness of companies to take up hovercraft development, HDL had to make things rather more attractive to enable them to take a bigger risk.

In the year following the SRN1 launching over four hundred inquiries were received by HDL from companies and organisations. The building of a manned prototype clearly had a major effect in stimulating interest in both the civil and defence fields. Eventually in 1960 four companies were granted licences by HDL in return for an advance against royalties. All built prototypes with aid from HDL (see chapter 15). They were Saunders-Roe (which had been taken over by Westland Aircraft), Vickers, Denny and Folland, a subsidiary of Hawker Siddeley. The industry was to be a 'balanced' one with Westland and Vickers concentrating on amphibious craft, Denny on rigid sidewall craft and Folland on overland applications (Vickers did in fact also produce an experimental Land

Rover to which skirts were attached, the idea being to use the air cushion to ease movement over rough and boggy land. This vehicle however formed a minor part of its total air cushion activities.) A Hovercraft Policy Committee was formed consisting of senior representatives of HDL and of the four firms to co-ordinate the industry's development and to discuss their problems 'with a view to establishing amongst themselves an agreed policy regarding the future development of hovercraft'.[3] This committee might well have developed into an organ for restricting entry into the industry and for ensuring that member firms pursued common policies in relation to market shares etc. However it is clear that it never developed in this way, partly because of the early emergence of Westland as the industry's leader. The Committee did however have a technical sub-committee which was largely responsible for drawing up draft regulations for the operation of hovercraft. NRDC saw the industry developing in an orderly fashion with HDL at the centre holding the key patents and licensing and assisting firms who could exploit the different applications of the air cushion principle.

The reasons that led these particular companies to enter hovercraft development varied, although there is an underlying thread: all were seeking to counter a potential or actual drop in sales. Westland had taken over Saunders-Roe principally for the latter's helicopter interests at Eastleigh. (This acquisition made their monopoly of British helicopter production complete.) Faced with Saunders-Roe's existing interest in hovercraft, and the poor alternative employment prospects on the Isle of Wight if this company's labour force was reduced,[4] Westland decided to continue. The financial commitment to hovercraft was not great in any case at that stage. The SRN1 programme was entirely HDL-financed, and the costs of the SRN2, a second generation craft, were to be shared with HDL. Denny, a small Clydeside shipbuilder with a strong scientific tradition,[5] had been experiencing a decline in orders for its conventional ferries and saw sidewall hovercraft as a possible answer. It also had its own testing tanks – an asset that was particularly useful for experimental work on hovercraft. It formed its own subsidiary company, Denny Hovercraft, to administer its new interests.

Vickers and Hawker Siddeley, like Saunders-Roe, had been affected by the setbacks in the defence programme. Vickers had suffered particularly badly as table 14.1 shows.

Table 14.1
Vickers: sales (£m.) and profits

	1956	1957	1958	1959
Shipbuilding	24	40	31	34
Engineering	36	48	43	47
Steel	33	38	35	33
Aircraft	65	70	65	41
Total	158	196	174	155
Return on capital (%)	13.5	12.5	10	7.5

Source: *Economist* (21 May 1960) p. 809

The South Marston factory of Vickers Supermarine had been excluded from the rationalisation of the industry in 1957. Its military aircraft production was therefore being run down and efforts were being made by the management to diversify into general engineering to provide continued employment for its design and production staff. Nuclear research and components, post office mechanisation and accelerated freeze drying (an unsuccessful project which was also supported by NRDC) were all fields in which the South Marston works put some effort during the late 1950s. At this time the local management heard about hovercraft and became enthusiastic about its potential (its design team had already had experience with flying boats). The main Vickers management, however, was much less keen. Hawker Siddeley's Folland factory at Hamble near Southampton was also facing the need to diversify, as its defence subcontracting work was drying up.

14.1.2 Withdrawal of Hawker Siddeley, Denny and Vickers

During the 1960s however the original pattern envisaged by NRDC was rendered impossible by a number of factors. Firstly, both Hawker Siddeley and Denny left the stable very early in hovercraft development. The first company stayed only two years. Its withdrawal may have been due in part to the fact that the 'sharp edges' of the 1957 White Paper were eroded in the early 1960s. Owing to the financial difficulties of aircraft companies in 1959–61, the government undertook to provide increased assistance towards producing new products, one of which was the Hawker Siddeley Trident.[6] New business was also coming in from elsewhere.[7] Then in 1962 the company's chairman stated: 'Quite candidly we are so wrapped up with our own aircraft development . . . we do not consider it right at the moment to go in for R & D of hovercraft, but we are interested'.⁸ Denny Hovercraft virtually ceased its activities when its parent company was forced into voluntary liquidation in 1963, following the continued contraction in its order book for military and ferry craft. Difficulties with the propulsion of the sidewall hovercraft had added further strain to the finances of the main company; over £400,000 in total had been spent on hovercraft development.⁹ The liquidators had originally tried to continue work on the hovercraft side but in the end this proved impossible.¹⁰

Vickers also eventually terminated its hovercraft interests. The main Vickers management started to lose what interest it had in hovercraft in the early 1960s, and had placed an embargo on the development of larger craft after the VA2, which was the last Vickers craft to be launched (in 1962).* There were also signs that Westland was beginning to emerge as the industry's leader. In 1963 the latter had received an order from the Ministry of Aviation for the conversion of its second SRN2 – which was then being built – into the military SRN3. Vickers had received no similar orders. The Westland management had also decided at some risk to lay down a production line for its new SRN5. Because of financial stringency, Vickers was unable to match this move. Furthermore, Westland had also taken the view that Vickers was infringing its patents. Although Vickers disputed Westland's claims (and took Queen's Counsel's advice, which was favourable to it), this was nevertheless another factor that accentuated the already lagging interest of the main Vickers management. In 1966 Vickers made its first step towards withdrawal when it joined with Westland and NRDC to form the British Hovercraft Corporation. The equity shareholdings were: Westland, 65 per cent; Vickers, 25 per cent; and NRDC, 10 per cent.

The merger of Vickers and Westland was formally justified by the companies and NRDC as being in the industry's best interests. In a joint statement the companies argued that 'this integration will avoid the inevitable duplication of research and development effort which has existed to date. British technical resources, facilities and finances must be concentrated to accelerate the creation of world markets for this new form of transport'.¹¹ The NRDC supported the ration-

---

* The VA2, although the last craft to be launched by Vickers, was smaller than the VA3.

alisation but reserved the right to license and give support to other applicants 'if the public interest so requires'. The merger took place in a period when the government was particularly sympathetic to mergers. The merger meant that in 1966 the industry was monopolised by one producer which in turn was dominated by Westland. (The newly formed BHC also acquired a minority interest in the Britten Norman hovercraft subsidiary, Cushioncraft, in the same year.)

In a static sense, the merger had several advantages. Vickers was moving along broadly similar lines of technical and product development. Furthermore, the market was not developing as originally envisaged. Concentration of resources on limited markets therefore seemed logically sound. It is much more difficult however to assess how far the merger might have affected the pace of technological development or the industry's willingness to identify and exploit new markets. Figure 13.1 may give some support to the hypothesis that, as Westlands emerged as the dominant firm, so its inventive activity declined, but the evidence is only very tentative.

In 1970, Vickers finally withdrew from the hovercraft industry, by selling its 1,250,000 £1 shares in BHC for 15p each. To Vickers the prospect of a return for its investment was becoming increasingly remote especially as the Westland shares had a prior claim on the profits. (BHC's R & D since 1966 had been financed by loans from Westland.) NRDC also withdrew in 1972, selling its shares for fifteen per cent of their face value.

BHC's rate of return (on an historic cost basis) in each year from 1967 to 1972 is given in table 14.2, which also provides data on other companies. Great care should be taken in making detailed comparisons between the years or companies because of different accounting conventions used. The figures are likely however to reflect broad magnitudes. Manufacturing industry as a whole earned about thirteen per cent rate of return on capital (on an historic cost basis) in 1969.[12] It is only in the last two years that the company earned returns around this level. In 1970, following three years of negligible returns, the company made a heavy loss owing 'to heavy development expenditure and service support mainly on Mountbatten [SRN4] craft and to substantial provision made against craft partly built in expectation of orders which in the Board's view should materialise but much later than expected'.[13]

THE DEVELOPMENT OF THE INDUSTRY 271

Table 14.2
Profits (Losses): hovercraft manufacturing
(£000s)

	BHC	Hovermarine Transport	Air Vehicles	Air Cushion Equipment
1967				
Profit (Loss)	109.2			
Net assets	6566.1			
$\frac{\text{Profit (Loss)}}{\text{Net assets}} \times 100$	1.6			
1968				
Profit (Loss)	43.9			
Net assets	4681.4			
$\frac{\text{Profit (Loss)}}{\text{Net assets}} \times 100$	0.9			
1969				
Profit (Loss)	23.4			
Net assets	7088.8			
$\frac{\text{Profit (Loss)}}{\text{Net assets}} \times 100$	0.3			
1970				
Profit (Loss)	(232.3)			
Net assets	7092.1			
$\frac{\text{Profit (Loss)}}{\text{Net assets}} \times 100$	−32.7			
1971				
Profit (Loss)	844.5	(33.6)	2.7	11.0
Net assets	6836.7	252.7	2.5	23.7
$\frac{\text{Profit (Loss)}}{\text{Net assets}} \times 100$	12.3	−13.3	107.9	46.5
1972*				
Profit (Loss)	916.2	98.7	5.2	15.2
Net assets	6279.1	294.2	13.9	43.3
$\frac{\text{Profit (Loss)}}{\text{Net assets}} \times 100$	14.5	33.5	37.0	35.2
1973				
Profit (Loss)				(3.5)
Net assets				57.0
$\frac{\text{Profit (Loss)} \times 100}{\text{Net assets}}$				−6.3

* At the time of writing 1972 was the latest year for which data were available for all companies except Air Cushion Equipment.

Note: Profits are taken before tax. Net assets consist of fixed assets *plus* current assets *minus* current liabilities. The average of net assets at

14.1.3 *New entrants*

The second factor that upset the concept of a planned industry was the entry of other firms. Some were already established in adjacent industries, while others were formed from scratch.

Of the established companies, the first to enter was a small firm in the Isle of Wight, Britten-Norman, which produced its own 'cushioncraft', CC1, without a patent licence from HDL. This company had been formed in 1954 by two aeronautical engineers who were providing aircraft maintenance and repair services and who specialised in the production and installation of crop-spraying equipment. Like Denny, they had a good technical record for their size. For example, by 1960 over two hundred aircraft throughout the world had been fitted with the Britten-Norman 'Micronair' rotary atomiser.[14] Their subsequent decision to move into light aircraft production in 1963 also showed sound technical and commercial judgement despite later difficulties. Sales of the company's Islander have been good.[15]

Britten-Norman entered the hovercraft industry following a request from Elder and Fyffes, the owners of banana plantations, who had examined the possibilities of using hovercraft for a quicker and smoother transport for their fragile and perishable fruit.[16] Britten-Norman's own interest in hovercraft had also been stimulated by their awareness of the shortcomings of surface transport for crop-spraying.[17] The CC1 craft was financed jointly by Britten-Norman and Elder and Fyffes. The latter company never took delivery of the machine, however, despite the earlier excessive optimism of its managing director, who, in 1960 could foresee 'all our produce being carried in this way in a couple of years'.[18]

---

the beginning and end of the year is used in the calculations.

Asset valuations appear to be on an historic cost basis in all four companies. However, in the case of BHC there have been changes over the period considered in methods of depreciation and valuation of stocks. These are unlikely however to affect seriously the comparability of the figures.

It should be remembered that hovercraft work is only one part of the activities of Vosper Thornycroft and Enfield Marine. Analysis of their accounts would not therefore be very meaningful in the context of hovercraft. And, as noted in the text, neither has sold any craft yet in any case. No published accounts showing any sales are available for Sealand Hovercraft. Air Vehicles profits have not been derived from sales of craft but from other hovercraft work.

Presumably the performance of the craft was insufficiently attractive. Nevertheless the craft produced by Britten-Norman, and later, its subsidiary Cushioncraft (the first craft to follow the SRN1) proved to be of considerable technical interest to government and other bodies, who appeared to have confidence in Britten-Norman's technical expertise: two of the second generation craft, the CC2, were ordered by the government, and the third was used in a joint development programme with HDL. The CC4 and CC5 were built as joint ventures with HDL and the CC7 prototype was bought by the government. HDL granted a licence to Britten-Norman soon after the CC1 was produced.

In 1970 however the Britten-Norman parent company found itself with Islanders on its hands following the collapse of the North American light aircraft market. Letters of intent to purchase several hundred Islanders were cancelled, the government refused to inject further funds into the company[19] – it had already provided launching aid for the Islander – and the company was eventually forced into liquidation. Cushioncraft had a cumulative adverse balance on revenue account of £448,000 by mid-1970. BHC took the remaining eighty per cent of Cushioncraft's equity. The CC7, the latest craft to be produced by Cushioncraft, is now being produced and sold by BHC. The takeover by BHC is of interest because the CC7 uses a skirt essentially similar to the HDL one. Hence BHC now have an operational craft that employs the basic system of their technical rivals.

The two other established companies to enter the field both had their main interests in marine craft. The first, Vosper Thornycroft, located near to Westland on the South Coast, had a long tradition in the production of fast naval vessels and was seeking to extend its interests in fast marine craft. It considered both hovercraft and hydrofoils. It eventually decided on hovercraft because its amphibious nature was attractive and because the company considered it was more likely to receive government help in view of the invention's nationality. Furthermore, 'there seemed to be special contributions an advanced shipbuilding organisation could make to hovercraft development, a field so far dominated by the aircraft industry'.[20] The company sought an unrestricted licence from HDL in 1965, and eventually obtained one in 1967.

The second company, Enfield Marine, specialised mainly in power boats. It launched its first hovercraft, the EM2, the first purpose-built freighter to be constructed, in 1971. (The EM1 was a design study only.) Neither company has yet achieved any sales. Vosper Thornycroft's only venture into the commercial operation of hovercraft proved commercially unattractive. It formed a joint operating company with a Swedish partner to run services between Malmo and Copenhagen, a route that was already served by boat and hydrofoil operators. Established operators on this route retaliated to the introduction of the new hovercraft service with its lower fares by reducing their own charges below the latter. The company's latest accounts allow £270,000 for losses arising from this venture that are deemed irrecoverable. Vosper Thornycroft has since decided to concentrate primarily on military applications of hovercraft.

The first important *new* company to enter the industry, Hovermarine, had been formed in 1965 by a group of men who had fairly close contact with sidewall hovercraft. Two, Ted Tattersall and David Nicholas, had been working on sidewall investigations at HDL (this side of HDL's activities had been run down after the collapse of Denny). Another founder, Hilary Watson, had originally been with Denny Hovercraft and was employed by the parent company's liquidators to continue work on the D2. Each knew the other well, and with a few others at HDL had engaged in an informal study group to discuss sidewall craft. The other two founders, Norman Piper and Jerry Hodgson, had earlier formed a company to market the Denny hovercraft although the liquidation had since made this objective redundant. 'Almost by coincidence', Tattersall met these two people at a meeting at HDL and found that they were interested in forming a joint venture. The company eventually (in 1967) obtained a licence from HDL to build a rigid sidewall craft.

The company has had a chequered experience. It was forced into liquidation in 1969. By March 1968 its current assets were less than twenty per cent of its current liabilities (table 14.3). When the liquidation was finally completed in 1972 creditors received only 5.9p in the pound. The company was eventually bought by a small new American company specialising in applications of the air cushion principle, and renamed Hovermarine Transport. The reasons for Hovermarine's failure probably stemmed from over-ambitious objectives at the start and the too early marketing of their craft before technical teething troubles had been eliminated.

THE DEVELOPMENT OF THE INDUSTRY 275

Table 14.3
Current assets over current liabilities

	BHC	Hover-marine	Hover-marine Transport	Sealand Hover-craft	Air Vehicles	Air Cushion Equipment	All quoted companies
1967	17.6	n.a.					1.8
1968	5.5	0.19					1.7
1969	4.7						1.6
1970	3.8				0.78	3.01	
1971	4.6		2.63	0.45	2.03	2.10	
1972	2.6		1.86	0.19	2.28	2.16	

Sources: Company accounts; *Annual Abstract of Statistics* London, HMSO (1971) table 384

(The problems faced by Hovermarine are dealt with more fully below.) The American company set its first objective as the elimination of all the craft's technical bugs. It has now done this and the craft is beginning to sell. This is reflected in its results for 1972 (see table 14.2). As a result of the original company's mistakes, therefore, one of the three major companies in hovercraft is now American-owned. Attempts by the directors to raise funds in this country to avoid liquidation proved unsuccessful. Unlike their future US parents, they were prohibited from raising money via a public subscription.

Several other newly established companies have followed Hovermarine into the industry. Hoverair was founded in 1966 by a hovercraft enthusiast, Lord Brassey, to develop and produce small 'fun' hovercraft under licence from HDL and BHC. In 1971, however, the company went into liquidation following the non-realisation of commercial orders, which at one time had been estimated to run into several million pounds.[21] Its net losses for its three and a half years of operation amounted to £155,000. Air Vehicles was established in 1969 by the former chief designer of Cushioncraft, and before that of Britten-Norman, Peter Winter, and by the managing director of Hovertravel, who at one time had also worked for Cushioncraft.

The objective of the company was to produce a small five- to six-seater hovercraft which was reasonably cheap to build and maintain. Winter left Cushioncraft because the parent firm had become so absorbed with its aircraft business that hovercraft development had in his view been given too few resources. At present (1974) the company's craft AV2 is at the prototype stage. Although during the last two years the company's rate of return on government and other contracts has been extremely high (table 14.2), it is clear that the company will require a new injection of capital if it is to bridge the gap between development and sales. The company has received an HDL advance (of £16,712 by the end of 1971–72) which considerably eased its liquidity position (table 14.3), although seventy per cent of its current assets in 1972 consisted of its prototype craft. The importance of this HDL loan can be gauged from the fact that it accounted for seventy-eight per cent of all sources of funds at the end of 1971–72. Although the company faces financial problems, enthusiasm for the craft and its potential among its small staff is very evident.

Sealand Hovercraft was formed by two friends who ran a small car repair and sales business. They had been approached by a common friend who asked them if they wanted to participate in building a hovercraft in their spare time. Their business grew from there; eventually the hovercraft project became so big that the motor side was run down. Like Air Vehicles, the company's objective is to build a small unsophisticated hovercraft. Unlike Air Vehicles, however, this company was formed by two men whose previous technical experience of hovercraft was minimal. At the end of 1971–72 the company was in severe financial difficulties, and its liquidity position was critical (table 14.3). Eighty per cent of its current liabilities consisted of a bank overdraft. It has received no public funds, a feature that is a crucial distinction between its position and Air Vehicles'. During the first part of 1973 the company managed to survive by financing all expenditure out of current sales. Again, despite these difficulties, the employees' enthusiasm for the craft and loyalty to the management is clearly evident (even though some regard this enthusiasm as misplaced): when the company ran out of money in November 1972 the labour force staged a 'work in'.[22] It is instructive to note that Air Vehicles, Sealand Hovercraft and Enfield Marine are all producing craft types that have not been attempted by BHC, although the latter has attempted to *modify* some of their existing craft, eg for freight purposes. It is too early to say yet whether these new craft will be successful. However, if they are it will provide support for the contention that commercial opportunities that are missed or ignored (sometimes for good reason) by a larger firm are often exploited by small 'outsiders'.

Only the currently more important companies have been mentioned above. There are however a number of other small companies, such as Air Bearings, developing or producing small special purpose or leisure hovercraft.

On the fringe of the hovercraft industry, there are a few small companies applying the air cushion principle to other uses. (BHC also has some, though not an extensive, interest in this field.) The most important of these 'fringe' companies is Air Cushion Equipment (ACE), which was formed by an ex-HDL employee, Les Hopkins, in 1968. Like Tattersall and Winter, Hopkins had never been in business before by himself. Like them too, enthusiasm to get his own ideas and those of others applied was a major factor in the formation of the company. The latter's commercial progress took him by surprise. As he says:

> ... the whole point of starting the company was that I had several industrial applications which no one wanted to take up. This (ie forming the company) was the only way I could get them out of my head. I would have been happy to have pottered along but the whole thing grew much more quickly than I anticipated.

The commercial basis for the new venture was a 'feeling' that there might be a market for trailers. The company's activities are based on the industrial application of the air cushion principle using HDL skirts, particularly in the movement of heavy loads. Since 1968 ACE has linked up with other companies in 'joint venture' programmes. It has half the equity of Mackley-Ace, a company constructing air cushion platforms for dredgers. It has also linked up with a much larger firm, United Builders' Merchants, to form a company, Hovertrailers International, to produce and sell hovertrailers. ACE owns thirty-seven per cent of the equity in this company. By collaborating in this way ACE has not had to finance the building of steel structures, but only the skirts for them. In 1970–71 and 1971–72 it earned high profits (table 14.2) and its liquidity position has been sound (table 14.3). Much of its success has been due to the fact that the company has been utilising existing technology and has not had its capital tied up in prototypes. Nevertheless it is in a vulnerable position: its financial results for 1972–73 show a small loss. ACE has financed itself largely from increasing its non-voting ordinary shareholdings (fifty-two per cent of its source of funds at the end of 1971–72).

278   CASE STUDY : DEVELOPMENT OF THE HOVERCRAFT

Several consultancies have also been set up by people who have had experience in either the production or the operation of hovercraft.

14.1.4 *Reasons for entry*

It is clear from the above discussion that the companies that have become involved in hovercraft in this country have done so for a wide variety of reasons. However, they may be grouped roughly into two main categories. Firstly, there are the 'defensive' innovators, who would probably not have become involved if their existing sales had not been threatened in some way. Vickers, Hawker Siddeley and Denny had all been affected, to different degrees, by a decline in their market prospects for their principal products and saw hovercraft work as a diversification which might provide new opportunities for growth and for the utilisation of existing capacity. Westland's involvement seems to have started in a much more accidental way, but, confronted with Saunders-Roe's interest in hovercraft and the threat of redundancies, it decided to continue the work. Secondly, there are the 'active' innovators : those who see an opportunity or have a technical idea and go ahead. Both Vosper Thornycroft's and Enfield Marine's entries were not due to a decline in the sales of their major products, but were more a logical extension of their existing marine interests. Hoverair, Hovermarine, Air Vehicles and ACE were formed primarily to exploit a particular technique or concept which in the founder's view was not being sufficiently utilised. All were formed by people who had been closely involved with hovercraft development, or had strong interests in it. Sealand Hovercraft was formed by 'outsiders' to exploit what they saw as a good market opportunity. Britten-Norman developed its CC1 following a request from a potential user, and after seeing potential in the air cushion for crop-spraying purposes. None of these new companies did any formal market research before embarking on development. It is probably true to say that some of these companies at least were production- rather than market-orientated; ie, they started from a technical idea which they had developed and wanted to see exploited commercially, without any clear conception of the size and nature of the market for which their products might be suitable.

It would be interesting to speculate on what would have

happened if Saunders-Roe had had no helicopter interests to make them attractive to Westland; or if demand for established products at Vickers, Supermarine and Denny had not been contracting. It is likely, given the later emergence of the new companies, that development would have gone ahead in any case, although in a different form and at a different pace. Hovermarine, it will be remembered, was formed in part from the frustration caused by the Denny collapse; ACE and Air Vehicles were also established because of the founders' frustration that certain markets and ideas were not being exploited.

14.1.5 The location of the industry

The industry is located mainly on the South Coast (the only major exception being Sealand Hovercraft, in Millom, Cumberland). The initial reason for this location was almost accidental – the existence of spare technical capacity at Saunders-Roe, which was used in the early days to examine the feasibility of the air cushion principle. (The first contract might easily have gone to Belfast.) The siting of HDL on Southampton Water then followed logically from the continuation of the Saunders-Roe hovercraft work on the SRN1. The industry grew from this base. Folland was located at Hamble and Vickers eventually undertook a lot of their hovercraft work in Southampton (Denny however was on the Clyde). Britten-Norman's and later Vosper Thornycroft's and Enfield Marine's interests were probably all stimulated by their physical proximity to the centre of hovercraft R & D and production. ACE, Air Vehicles and Hovermarine were direct offspring from existing organisations in the area and there were sound reasons for their being similarly located. Southampton Water also offers good facilities for testing hovercraft. (These facilities were a major factor in the original location of the flying boat experiments.)

14.1.6 The financial problems of companies

One of the characteristics of the industry's history has been the number of companies that have failed or gone through severe financial difficulties. Denny Hovercraft, Hovermarine and Hoverair have gone into liquidation. Cushioncraft was taken over by BHC after the failure of the parent company. Sealand Hovercraft nearly went out of business in 1972. Air Vehicles' position is certainly not strong. This record in the industry may of course be a reflection of the degree to which the hovercraft is a marginal innovation, at least as it has been translated into the particular products of these companies. However, there may be special financial problems facing companies developing a new technology that is basically viable. Some of the demand and supply factors affecting financial needs are examined in turn below. These two factors are closely interlinked – for example, more finance may be needed to modify a craft in order to increase its sales prospects; increased sales may in turn provide greater availability of funds in the future through higher profits. However, for the sake of exposition, the demand and supply factors are separated in the following discussion.

Demand

We are not concerned here with a general analysis of the demand for finance, but with those aspects relating to the exploitation of a new technology, which are likely to require financial resources additional to those needed by a firm manufacturing a well established product.

Firstly, the length of development and financial outlay before achieving any sales revenue can be considerable. It may take two years or more to design and build a craft, depending on the type of craft involved. Production trials also have to be undertaken, and certification obtained, before marketing can be started. Both may give rise to extensive modifications. Production facilities then have to be built; these will probably also have to be modified in the light of experience.

Secondly, in using any new technology it is likely that unforeseen problems will be encountered which in turn will raise costs and usually lengthen the period of modification. An example of the difference between the estimated and actual costs of constructing a prototype craft in one company in the industry is given in table 14.4. In another company, the final cost of a prototype was 100 per cent higher than the estimate made before construction. These increases may seriously affect the financial standing of the company especially if it obtained external finance on the basis of its cost estimates. The increases in table 14.4 reflect the fact that many aspects of hovercraft design, construction and operation were entirely new to the firm. Under-estimates may of course be made

Table 14.4
An example of increasing costs during prototype construction

Item	% increase in manhours	% increase in expenditure
Technical department		
Up to launch	171	
Trials	(from nil to 22,000 hrs)	
Drawing office		
Up to launch	400	
Modification	50	
Construction Department		
Up to launch	227	
Modifications, etc	246	
Bought out materials, equipment and services		14
Certification		120

Table 14.5
Increases in estimated operating costs between date of launch and date ready for commercial service

Item	% increase in costs per mile*
First cost	167
Insurance	200
Crew	100
Engines	18
Skirts	233
Propellers	—
Fuel oil	25
Indirect	38
Miscellaneous & labour	1300

* The same load factors, total operating hours, speed and sea states are assumed.

deliberately (or subconsciously) where estimated cost figures are likely to play a crucial part in deciding whether a project should go ahead or whether finance will be made available from external sources. Under-estimates of operating costs may be made for similar reasons (see below).

Thirdly, and linked with the second point, actual operating costs of a craft may be much higher than anticipated because of unforeseen technical problems. Table 14.5 gives an example of the extent to which operating costs may have to be revised during the trials period – in the example a period of just over a year. The figures relate to one of the larger hovercraft.

A small element of these increases is due to inflation; however, the majority of the change again stems from the newness of the technology and the consequent underestimation. It is interesting to note too, that the biggest single increase in the estimated costs, excluding the 'Miscellaneous and Labour' category, is with skirts with which the company's designers and engineers were least familiar. The massive increase in 'Miscellaneous and Labour' costs was primarily due to the realisation that maintenance labour was likely to be significant. In the early estimates labour charges were assumed to be included within the cost of each component. The company concerned found however, after initial trials, that it was not realistic to apportion charges in this way and that there would be a standing labour charge necessary to deal with inspections, adjustments, anti-corrosion etc, quite unrelated to the replacement of defective components. The company is now directing a good deal of its development effort to reducing these labour costs. Increases in operating costs may seriously affect sales prospects in particular markets and may therefore require further development finance aimed at reducing costs.

Underestimates of this magnitude can severely strain the finances of a small and relatively new single-product company which has no record of technical or commercial success. It is probably also true to say that the company that is relatively new and whose staff have only limited business experience is particularly likely to make the bigger under-estimates of costs. Enthusiasm for the project may disguise the seriousness of the cash flow problem. For example, Hovermarine started design work on its larger craft, the HM4, before the first HM2 was even produced. As a consequence of this kind of enthusiasm, new projects grow and old ones are extended, overheads increase and the labour force builds up. In this respect it is interesting to compare the employment patterns of Sealand Hovercraft and Air Vehicles, two companies each with one main project aimed at roughly similar markets. Sealand Hovercraft started marketing its hovercraft in 1972 and employed a (maximum) labour force of eighty-eight in that year. Air Vehicles, whose craft is still at the prototype stage, employs six workers. The very considerable difference between the size of the two companies cannot be explained solely by the different stages of the development–production–marketing spectrum at which the companies are working. Hovermarine's growth was also rapid considering the stage it was at: by 1968 it was employing eighty personnel. Westland on the other hand has kept its development strategy under close control. In 1965 the Plowden Report on the aircraft industry noted that the company which – probably by no coincidence – was one of the most consistently and highly profitable in the industry had avoided research and development risks that would strain their finances; and :

> ... they have rarely had to incorporate in any new venture more than one major new development. New types they have taken on have had either a new engine installation, a new air frame or a new major system, but not normally a combination of these.[23]

This cautiousness is apparent in BHC's contribution to the technical development of hovercraft and in its current policy towards them (see p. 262).

The experience of Hovermarine is a good example of a company with what appears to be a basically sound technical and commercial concept and with plenty of enthusiasm in its staff, but which failed to institute any effective control system on its expenditure. Had it kept a rigorous control over its design operations in the early years, the original company might still be in business. Hovermarine's history poses a fundamental problem in relation to the innovation process. How can technical enthusiasm and ability be kept in balance with sound financial control? Control at arm's length by outside financiers does not seem to provide the answer. What seems to be needed, particularly in the case of the smaller company, is participation by advisers in the company's day-to-day management. This is a burden however that most banks and other financial bodies would not be prepared to accept, nor, in many cases, would the companies involved. If this is the case, then financial failure seems an inevitable part of the innovation process.

Supply

Whether internal sources of finance will be sufficient for an established company to finance a new development will depend on numerous factors, including the size and complexity of the development and the profitability record of its parent firm. In the case of the new firm, the resources of the founder and his own reputation in commercial and technical terms will be important. BHC has obtained most of its finance from a parent company which as shown above has had a good record of profitability. It also has a fairly diversified product range. Both Denny Hovercraft and Cushioncraft, on the other hand, were closely tied to the fortunes of two fairly small parent companies, both of which had only a narrow product base, and whose profitability record was certainly not outstanding. The company that is entirely new may be in an even more difficult position.

Faced with limited financial funds a company may attempt to market its product before all the technical 'bugs' have been removed. This in turn creates further problems and the expense of repairing and maintaining craft in service. Premature marketing may also reflect adversely on future sales. Too early marketing of the HM2 was another factor in Hovermarine's demise. As shown later, the HM2 was very unreliable when it first went into service. (The marketing operation itself may substantially increase a new company's demand for financial resources at a critical time. This is particularly so in hovercraft operation, where a worldwide network seems to be essential). A new company may of course try to diversify, financing its major development out of profits from other sources. For example, Air Vehicles financed its early work on its craft partly out of profits from small government contracts and from work carried out for Hovertravel's subsidiary, Hoverwork. (The managing director of Hoverwork is also a director of Air Vehicles.)

However, where internal sources do not provide the necessary capital, the founders of a new firm may attempt a number of possible approaches to raise capital. It may try to raise additional share capital from friends and acquaintances and from institutions. This may however have the effect of diluting the founders' controlling interest, which they may be very reluctant to do, although this problem may be partly avoided if the company can get someone to take up shares without voting rights. Hovermarine raised £50,000 from Czarnikow, the commodity brokers, in exchange for preference shares and a small stake in the company's equity. It later also obtained £500,000 from William Cory, the shipping group, but this time a substantial part of the equity was given in exchange. (Czarnikow and Cory together accounted for forty-nine per cent of the equity at the time of liquidation.) Up to the time of the Czarnikow injection of capital the Hovermarine directors 'had been knocking on someone's door every day of every week but possible financiers generally wanted a bigger share of equity than we were prepared to let go'. The company first came into contact with

Czarnikow through an advertisement in a paper. The personal assistant to the Czarnikow chairman 'showed obvious enthusiasm and it was his influence which led the chairman to do a further investigation of us'.[24]

It is doubtful whether the capital would have been available but for the personal assistant's influence and the chairman's own enthusiasm, again illustrating the importance of personal factors. Hovermarine attempted to raise further institutional finance in this country at the time of liquidation, but was unable to do so. It had already attempted to market its craft and the latter's technical troubles did not make the company sufficiently attractive to possible backers.

Sealand Hovercraft had obtained some additional finance by bringing in new partners but its principal source of funds (up to 1972) has come from its bank overdraft. This of course has severely strained its liquidity position. The company has received no government funds specifically for its hovercraft work apart from a prototype grant for its SH1. (It has however received regional assistance, of which the prototype grant was one kind.) Before September 1973 it had been unsuccessful in raising additional funds in industry, but without access to correspondence it is impossible to be objective in analysing the reasons why such finance has not been forthcoming.

Air Vehicles has obtained some of its finance from HDL. Consequently it is able to pursue its work without diluting ownership control and without straining its liquidity position to the same extent as Sealand Hovercraft.

It is difficult with the information currently available to assess whether there have been any shortcomings in financial markets. The Economists' Advisory Group studied the availability of finance for innovation in its report on *Financial Facilitites for Small Firms* for the Bolton Committee.[25] The impression the researchers gained from interviews with NRDC, TDC and other institutions 'was that someone with a good product and good management potential would have no problem in obtaining finance'.[26] Again, it is too early to assess which are 'good products' in the hovercraft industry. Denny and Hovermarine however were (successively) precursors of the present Hovermarine Transport. If the sales of the latter's craft continue to rise, the experience of the two earlier companies may indicate some shortcomings in the finance market. In the case of Hovermarine, poor management of funds up to the time of liquidation was clearly one factor in reducing the availability of external finance. If financial institutions, when examining the needs of a small firm, accept the latter's management *as it is*, then they will clearly often be at risk in providing funds. But perhaps more attention should be given to infusing good management, as well as additional finance, into a company.

14.1.7 Foreign producers

HDL and Vickers jointly licensed Mitsui in Japan in 1961 and HDL and Westland licensed Mitsubishi. Because of the United States' Jones ('Foreign Bottoms') Act, it was also decided to license Bell Aerospace (HDL and Westland) and Republic Aviation (HDL and Vickers). Republic Aviation went into liquidation shortly afterwards. Hovermarine Transport completed negotiations with HDL for a licence to produce in the United States in 1973. According to *Jane's Surface Skimmers 1972/3,* the countries with the greatest involvement in hovercraft production (as measured by the number of companies involved) are the US (twenty-four companies) and Canada (nine).

14.1.8 The effect of patents

Patent holdings have clearly had some influence on the industry's development. Although, as indicated in chapter 13, there is some dispute over who holds the important patents, no potential hovercraft producer has felt itself to be in a sufficiently strong position to go ahead without an HDL licence. One company took the view that, although it doubted whether HDL did in fact have the rights it made itself out to have, the blessing of a quasi-government body gave its activities a special standing and made it more eligible for aid from government departments: the latter would not have been given to any company that did not have a licence from HDL.

HDL's role as a body with some public accountability through NRDC makes its patent position somewhat more vulnerable than if it were a purely private body subject only to the conditions of the Patents Act. For example, the delay over the granting of Hovermarine's and Vosper Thornycroft's licence was raised in Parliament.[27] It would also be very difficult for such a body to refuse a licence to a firm that had produced a craft that was technically good and had some commercial prospects. Britten-Norman's acquisition of a licence *after* it had built CC1 is interesting in this respect. Refusal to grant a licence may however act as a disincentive to a firm to start work on a craft.

Although HDL's power is somewhat circumscribed in this way, it has nevertheless been able to impose certain limitations on markets in which sales can be made and on the craft sizes and types that can be produced by a licensee. For example, Britten-Norman was limited to producing craft below, and Vosper Thornycroft above, a given weight, and Hovermarine Transport is restricted to building sidewall craft.

HDL's patent policy is dealt with more fully in chapter 15. It is worth noting here however that its attitude seems to have been more restrictive to new entrants in the past than it currently is. For example, HDL's acceptance of Vosper Thornycroft's application for a licence was considerably delayed; initial discussions started in September 1965 and were concluded only in September 1967, because of objections by BHC to HDL's granting a (relatively) unrestricted licence to Vosper Thornycroft. Since 1966–67, however, several HDL licences have ben granted to companies. This easing may be due to the fact that in the mid-1960s it was becoming clear that HDL's income was not developing as anticipated.

It is probably true to say that in the long run the effect of patents on the industry's development has been marginal. No major company has been unable to obtain a licence, although it is not of course known whether any companies seriously *considered* hovercraft work but did not pursue it because of the existing patent situation. No company has received substantial royalty income. Vickers' exit may have partially resulted from BHC's patent claims but other factors were also involved. The existence of HDL's patent portfolio probably resulted in temporary *delays* for companies intending to enter the field, although it must be realised that in a more fundamental sense the existence of HDL's own portfolio offering an alternative skirt system considerably reduced the power of BHC itself to restrict entry.

Even when a firm has acquired a licence, however, it still has to develop a prototype to a marketable stage. This normally requires considerable finance, but also commercial and technical know-how that is not embodied in patent specifications. The experience of many companies involved in the industry has shown that these factors constitute a much more formidable barrier to entry than formal patent rights. Hovermarine's experience also highlights a special problem that may be faced by *new* companies in relation to patents. One of the original directors told the author that, in trying to raise finance for the company, they were caught in a 'chicken and egg' problem: possible financiers would not give help until an HDL licence had been granted; and HDL would not provide a licence until there was evidence of financial backing. The problem was only resolved when Czarnikow agreed to provide finance on the basis that a licence would be forthcoming.

14.2 Operators

Like hovercraft production, hovercraft operating activities have been marked by a high birth and death rate. Only three companies have operated for any length of time in Britain. Hoverlloyd was founded by two Swedish shipping companies, Swedish Lloyd and Swedish American, who wanted to enter the highly profitable cross-Channel ferry services. Because the company was unable to obtain docking facilities at Dover, owing to pressure on resources from existing operators, it decided to enter hovercraft operation using a different base. It was the first company to order an SRN4 in 1965 while the craft was still at the design stage. At first, it operated SRN6s to gain experience. The company now operates three SRN4s from Pegwell Bay, near Ramsgate, to Calais. It has no other services.

The company made continuous losses in the first few years of operation, as is shown in table 14.6, despite the fact that it was operating in competition with conventional operators who collectively maintained their fare levels (see below). The figures give some idea of the costs involved in operating in a new area of technology. Some of the reasons for these losses are outlined later.

British Rail entered hovercraft operation in 1966. It is difficult to ascertain how far this resulted from a political decision (taken in an environment that was becoming increasingly technology-conscious) or from purely commercial considerations. Certainly, the Ministry of Transport was having discussions with British Rail about the development of hovercraft services.[28] The Ministry of Technology was also asking BR (in 1965) to 'continue its experiments' in hovercraft operation.[29] Like Hoverlloyd, British Rail's subsidiary, British Rail Hovercraft has made considerable losses (table 14.6).

British Rail Hovercraft have operated on the Solent and across the Channel (from Dover to Boulogne and also, after 1970, to

Table 14.6
Profit (Losses): hovercraft operations
(£000s)

	British Rail Hovercraft	Hoverlloyd	Hovertravel
1967			
Profit (Loss)	(149.0)	(76.7)	12.0
Net assets	178.3	613.3*	81.4
$\frac{\text{Profit (Loss)}}{\text{Net assets}} \times 100$	−83.6	−12.5	14.8
1968			
Profit (Loss)	(232.3)	(45.7)	2.3
Net assets	1147.3	1548.2	75.0
$\frac{\text{Profit (Loss)}}{\text{Net assets}} \times 100$	−20.2	−2.9	3.2
1969			
Profit (Loss)	(287.9)	(504.8)	25.4
Net assets	2572.1	2975.5	94.1
$\frac{\text{Profit (Loss)}}{\text{Net assets}} \times 100$	−11.2	−16.9	27.0
1970			
Profit (Loss)	(421.1)	(512.0)	30.2
Net assets	2574.3	3296.1	107.3
$\frac{\text{Profit (Loss)}}{\text{Net assets}} \times 100$	−16.3	−15.5	28.1
1971			
Profit (Loss)	(198.8)	(90.7)	3.8
Net assets	1933.3	3086.5	114.2
$\frac{\text{Profit (Loss)}}{\text{Net assets}} \times 100$	−10.2	−2.9	3.3
1972			
Profit (Loss)	18.0	n.a.	(66.9)
Net assets	1830.1		92.8
$\frac{\text{Profit (Loss)}}{\text{Net assets}} \times 100$	−0.9		−72.1

* Value of assets at year end
Notes as for table 14.2, pp. 271–2

Calais). On the Solent, three services have been run at various times: Southampton to Cowes from 1966 to date; Portsmouth to Cowes (1967–69) and Portsmouth to Ryde (1968–72). The first two services used SRN6s and the third an HM2. The Portsmouth–Cowes service closed primarily because traffic failed to live up to expectations. The Portsmouth–Ryde service was discontinued because of mechanical problems and docking difficulties at Ryde pierhead.

The only private company to operate hovercraft for more than a brief period is the small Isle of Wight firm, Hovertravel. This company was formed by Britten and Norman (who later withdrew) and a few associates including a consulting engineer, an accountant and a stockbroker, who was particularly enthusiastic about hovercraft, and who had built his own craft. They managed to raise capital from friends and others. Most of the shareholdings are held by interested individuals rather than by institutions. This company, with its subsidiary Hoverwork, has managed to make a profit for most of the years that it has been in operation (table 14.6), although it made a very heavy loss in 1972 as a result of a drastic decline in the turnover of Hoverwork, its subsidiary, which specialises in activities where the hovercraft's amphibious capability is particularly suitable, eg seismic surveys, etc. This major reverse in Hovertravel's fortunes again illustrates the vulnerability of a small, narrowly based company to changes in market conditions.

Hovertravel and Hoverwork however have played an important part in hovercraft development which seems out of all proportion to their size and to their overall financial profitability. They have demonstrated the usefulness of the amphibious capability of hovercraft in certain situations and in various parts of the world; they were the first company to start a regular hovercraft operation which has since been maintained; and, for better or for worse, they drew up (in 1967) an outline specification for a craft based on their previous operating experience which was used as the basis for the VT1.

Both Hoverlloyd and British Rail Hovercraft are backed by large parent companies. At the end of 1971 Hoverlloyd owed £2.4m. to its parent companies which also owned £1.0m. in shares. British Rail Hovercraft owed £3.4m. to its nationalised parent at the end of 1972 (only a token shareholding has been paid up). All of Hovertravel's capital however comes from shareholdings.

The existence of a nationalised body (British Rail Hovercraft) competing with a private company (this has been especially true on the Solent routes) does raise the problem of the 'fairness' of competition. Hovertravel for example have argued (somewhat predictably) that British Rail Hovercraft's special financial position can enable it to continue to make losses whereas Hovertravel must pay its way.[30] Up to a certain point however, a large private company might be prepared to operate at a loss for a few years before making a profit. Hovertravel under its present financial structure could not sustain this.

Several other companies have also been involved in hovercraft operations. Vosper Thornycroft's service in Scandinavia has already been mentioned. In 1965 Clyde Hover Ferries, a subsidiary of a small Scottish company, Highland Engineering, was formed to operate services between the mainland and the Scottish Isles using an SRN6. (Its founder owned one of the islands off the Ayrshire coast.) Services were suspended in 1966. The company failed for a variety of reasons: traffic did not live up to expectations; there were strong objections to the noise generated by the craft by local residents; and the company was probably too small to manage the services adequately. Another service was started on the Clyde in 1970 by the Caledonian Steam Packet Co., part of the nationalised Scottish Transport Group which runs the steamer services on the Clyde, using an HM2, but the service lasted only a year. Again, traffic did not live up to expectations and mechanical problems lowered reliability, as shown by table 14.7. The company estimates that in the summer of 1971 alone it made a total loss of £25,000 (including depreciation) on the services.

In 1968 a group of local businessmen decided to set up a company, Humber Hoverservices Ltd, to run a service between Hull and

Table 14.7
Reliability of HM2 on Clyde, 1971

Maximum possible number of days in operation	112
Days completed with full schedules	55
Days on which no service ran due to weather	4
Days on which partial service ran due to weather	2
Days on which no service ran due to mechanical troubles	39
Days on which partial service ran due to mechanical troubles	6
Days on which services were curtailed because of staff problems	6

Note: Hovermarine's records give only 23 days on which the craft was not able to run due to mechanical troubles.
Source: Caledonian McBrayne

Grimsby using an HM2. However, this service also failed to last very long owing to weather and mechanical problems.[31] During the first three months of operation alone, eleven propellers and fourteen propeller shafts had to be changed. The most recent company to be established is the Southampton-based International Hoverservices, which was founded by one of the initial Hovermarine directors, Norman Piper. It operates only sidewall craft. It currently has a contract to ferry workers each weekday, morning and evening, between the Isle of Wight and Southampton and it runs a summer service between Bournemouth and Swanage. In addition, it is operating a service on the Thames under contract with the government.

A few other companies have also used hovercraft in this country, but have discontinued the services after a short interval, often going into liquidation as a result. Some, like the British United Airways service between Rhyl and Wallasey in 1962, were *intended* to be only experimental, although even here the experience was not encouraging : of the 57 days on which the services were scheduled, 32 saw no operations owing to weather and mechanical difficulties.[32] No major British private company therefore has entered hovercraft operation; the biggest private operator is a foreign company which started hovercraft services because access to other (conventional) operations was denied to them.

The reluctance of established companies to enter this field may be taken as a sign of lethargy and unwillingness to innovate. There is however little incentive for established ferry operators to enter an entirely new field, especially where existing services are earning satisfactory profits, as they are on the Channel, and where the economics of a new form of transport appears very doubtful. It must be remembered too that it will be profitable to replace existing equipment by an innovation only if the operating costs of the former exceed *both* the operating *and* capital costs of the latter (see p. 178). Hovercraft are almost certainly not meeting that condition. Furthermore, the experience of Hoverlloyd and British Rail Hovercraft has shown that there are substantial costs for innovating in transport. In the case of the SRN4 in particular, where the first craft went into service almost immediately after production, teething troubles have been extensive, and development work on skirts and engines especially has been necessary. This in turn has led to cancellation of services, as shown by table 14.8.

THE DEVELOPMENT OF THE INDUSTRY   293

Table 14.8
SRN4: flights cancelled, 1968, 1969, 1972

Year	Total flights timetabled	Skirt and associated problems	Engines	Other craft faults	Adverse weather	Total
1968 (1/8–13/10)	454	108 (23.8%)	18 (3.9%)	18 (3.9%)	58 (12.8%)	188 (41.4%)
1969 (2/4–31/12)	1868	242 (12.9%)	72 (3.8%)	72 (3.8%)	87 (4.7%)	452 (24.2%)
1972	6056	120 (2.0%)	91 (1.5%)	244 (2.7%)	542 (8.9%)	1169 (19.3%)*

*This total also includes 254 cancellations due to trips not being 'commercially viable' and to 'hazards'.
Source: British Rail Hovercraft;
Air Cushion Vehicles

The record of the SRN4 is clearly improving, but the costs of operation in the early years have been heavy. The N6s on the Solent route have steadily improved in their performance and cancellations are now very low. In 1972 total cancellations of all kinds amounted to only 1.64 per cent of all flights scheduled. However, this reduction of cancellations of itself has been insufficient to achieve economic operation. The HM2's reliability on the Solent, although initially very poor, also showed continuous improvement during the life of the Ryde–Portsmouth service. Despite the reluctance of established firms to enter the field, there has been no shortage of small companies willing to 'have a go'. (As we have seen, this has also been true in hovercraft production.) Most of them may be doomed to failure from the start, but they do at least enable possible new markets to be tested.

14.2.1 *Competition with other forms of transport*

The hovercraft has come into perhaps the fiercest commercial battle on the Southampton–Cowes route. Until the first hovercraft service was established in 1966, the ferry service was monopolised by the long-established Red Funnel Company which operated conventional ships. This company looked at the possibility of introducing the HM2 on the run but decided against it on the grounds of its poor reliability record. In 1969 it introduced a hydrofoil service in an attempt to curtail the market penetration achieved by the hovercraft service. Both the hovercraft and hydrofoil take twenty minutes compared with fifty minutes by boat. Table 14.9 gives some indication of the shares held by the three forms of transport of the 'unaccompanied' passenger market in the last few years.

Table 14.9
Market shares on the Southampton–Cowes route: foot passengers
(% of the total market)

Year	Ship*	Hydrofoil	Hovercraft
1966	95		5
1967	88		12
1968	85		15
1969	77	6	17
1970	75	7	18
1971	74	6	20
1972	75	6	19

* Because the ships also carry cars with accompanying passengers these figures are estimates only.
Source: Red Funnel;
British Rail Hovercraft

The table must be treated with care as the services offered are differentiated in a number of respects. Landing points at Cowes and Southampton for the hovercraft differ slightly from those for other services. Car parking facilities at the terminals also differ, and British Rail Hovercraft provides a 'free' bus service between the Southampton terminal and the railway station. Frequencies of the services also differ.

For the moment, the fast ferry services seem to have stabilised at around twenty-five per cent of the total foot passenger market, the hovercraft route having grown very rapidly in the first three years of its life. Those using the service are most likely to be commuters and business travellers rather than holidaymakers. The hydrofoil fares were slightly below the hovercraft fares until May 1972, when they both settled at the same level (100p per single journey compared with 55p single on the boat). In October 1972, however, Red Funnel cut its fare by 25p, and its hydrofoil fares have remained at a level twenty-five per cent lower than the hovercraft fares since that date. The company is clearly aiming at forcing British Rail Hovercraft out of the route and is using its profits from its vehicle ferry work to cross subsidise its hoverfoil services. The latter are losing heavily: in 1973 passenger traffic was only fifty-five per cent of that necessary to break even on direct operating costs.

British Rail has little room for manoeuvre to respond to this price cutting : its service is also operating at a considerable loss. It is likely however that *one* fast ferry operator could show an operating profit. If, for example, hovercraft passengers transferred to the hydrofoil, Red Funnel would more than break even. Thus competition in this case has led to a situation of cut-throat competition and unprofitable operation. Fast ferry passengers are benefiting at the expense of other types of passenger.

Competition between hovercraft and conventional ferries on the cross-Channel routes has been complicated by the existence of the Harmonisation Conference of which BR Hovercraft (but not Hoverlloyd) is a member. This Conference fixes the minimum fares for the main categories of traffic. In 1972 BR Hovercraft's fares were only slightly above this minimum. British Rail defended its membership of the Conference to the Monopolies Commission on the following grounds :

> . . . the accounts of the company [ie British Rail Hovercraft] showed that there was no basis for charging fares lower than those currently charged by ships. Indeed a case existed for charging fares greater than the current fares charged by ships and there had been attempts from time to time to do this. For the present however there was a sound commercial case for keeping hovercraft fares down to those of ships.
> . . . The commercial case for this policy rested upon
> (a) the necessity for quick and simple inter-availability between hovercraft and ships, thus encouraging voluntary transfer of ship passengers to hovercraft;
> (b) the encouragement of unbooked traffic to opt for travel by hovercraft; and
> (c) generally the necessity for this relatively new form of traffic to establish itself within the existing market as an acceptable means of crossing the Channel.[33]

This statement by BR Hovercraft seems to imply that its attempts to charge a higher fare, to reflect the higher operating costs of hovercraft, have been unsuccessful and that it has therefore decided to keep its prices in line with shipping fares. The arguments put forward in support of this latter policy need examination. Argument (c) may carry some force, but this must decline with time : hovercraft have already been operating for five years on this route. Argument (b) if distinct from (c) seems strange, bearing in mind that a different *type* of service with different costs is being offered.

Argument (a) was not accepted by the Monopolies Commission as sufficient justification for maintaining common fares.

The Monopolies Commission concluded that the price-fixing of the Conference was against the public interest, on the grounds that it may have inhibited the drive for increased efficiency (although they had no direct evidence on this score) and that it took only limited account of the varying levels of demand on different days and in different weeks. (The Conference provided only two sets of fares: 'off season' and 'standard'; there were however very wide variations in demand pressures in the period to which the latter fare was applicable.) The Monopolies Commission took the view that by restructuring fares within the summer peak period in an attempt to even out demand, the volume of resources used in the services could be 'materially reduced'.[34] It also argued that the existence of the Harmonisation Conference was likely to inhibit the emergence of a more rational price structure.

It will be interesting to see the effects of this recommendation on hovercraft travel. If the *general* level of costs of operating ship ferries fall as a result of increased efficiency, then this will accentuate the difficulties of BR Hovercraft, especially if demand conditions do not permit it to charge materially higher prices. It may be argued that BR Hovercraft may also increase its efficiency. However, because it is a relatively young organisation, the scope for this may be less than for an organisation that has lived under a price fixing agreement for over ten years. On the other hand, if the abandonment of price-fixing allows BR Hovercraft to have a more flexible pricing structure, then its revenue may well improve without any increase in costs. Hoverlloyd already has a more flexible price structure than its nationalised counterpart, although it must be remembered that, up to 1972 at least, it had not earned a positive rate of return (see p. 289).

NOTES on Chapter 14

1. W. Browning, General Manager of Saunders-Roe, quoted in *Flight* (January 1958)
2. J. Rapson *The Development of Hovercraft in the UK*, 2, Hythe (December 1970)
3. Chairman of HDL, quoted in *Flight* (August 1960)
4. In *The Times* (15 July 1959) it was reported that Westland had made

THE DEVELOPMENT OF THE INDUSTRY 297

it clear in their take-over statement 'that they were very conscious of the unemployment problem in the Isle of Wight . . . and hoped to increase rather than diminish their activities in that area'.
5. *New Scientist* (August 1961). See also *Shipping World* (30 September 1953)
6. *Report of the Committee of Inquiry into the Aircraft Industry* Cmnd. 2853 London, HMSO (1965) p. 17
7. *Economist* (9 June 1962)
8. Quoted in *Flight* (5 July 1962)
9. Statement by the liquidators: report in the *Glasgow Herald* (18 October 1965)
10. *Glasgow Herald* (18 October 1965)
11. *Hoveringcraft and Hydrofoil* (March 1966)
12. M. Panic and R. E. Close 'Profitability of British Manufacturing Industry' *Lloyds Bank Review* (July 1973)
13. D. C. Collins, Chairman of Westland Aircraft, quoted in *Hoverfoil News* (18 February 1971)
14. J. W. R. Taylor (ed.) *Jane's All the World's Aircraft 1960/61* London, Sampson, Low Marston (1961) p. 30
15. D. G. Lethbridge *The Islander — Problems of the Innovational Entrepreneur* Milan, International Institute for the Management of Technology (1972)
16. *Flight* (18 March 1960)
17. *Flight* (24 May 1962)
18. Quoted in *Flight* (17 March 1960)
19. Editorial, *Flight International* (4 November 1971)
20. C. Dawson *Quest for Speed at Sea* London, Hutchinson (1972) p. 81
21. *Hoverfoil News* (22 July 1971)
22. *Financial Times* (1 December 1972)
23. *Report of the Committee of Industry into the Aircraft Industry* Cmnd. 2853 London, HMSO (1966) p. 65
24. Interview with T. Tattersall
25. *Report of the Committee of Inquiry on Small Firms* Cmnd. 4811 London, HMSO (1971)
26. ibid. p. 54
27. *Hoveringcraft and Hydrofoil* (March 1968)
28. *Hansard* (Commons) Col. 209 (22 June 1965)
29. *Hansard* (Commons) Col. 208 (22 July 1965)
30. *Annual Report* (1968)
31. *Hull Daily Mail* (18 November 1969)
32. *Economist* (13 October 1962)
33. Monopolies Commission *Cross Channel Car Ferry Services* London, HMSO (1974) p. 30
34. ibid. p. 59

15 Public Involvement

Public finance for hovercraft development has come through two main sources: NRDC (and its subsidiary HDL) and government departments. The NRDC has a statutory obligation to ensure that it breaks even, 'taking one year with another' (p. 146), although it has interpreted this obligation very broadly. Government departments work under no such explicit statutory obligation. The NRDC is not, of course, an entirely independent body: it receives its funds from the Department of Trade and Industry (DTI) (now the Department of Industry) and is accountable to this department for its expenditure.

The first section of this chapter deals with the part played by NRDC and HDL; the second discusses the role of government departments.

15.1 NRDC and HDL

HDL was set up in 1959 to manage NRDC's hovercraft work. This followed from the anticipation that hovercraft support was likely to become the biggest single project assisted by the parent body and from the more immediate need to monitor the building of the SRN1. Its activities were two-fold; to manage the patent portfolio which was built up during its life and to run a technical group to support the industrial development of the air cushion principle. However in 1967, for reasons discussed in chapter 13, the Technical Group was transferred to the National Physical Laboratory (NPL). By 1972 this laboratory's interest in hovercraft had become nominal. HDL maintained only a skeleton staff after 1967 to manage the patent portfolio. By 1972 its cumulative deficit stood at just over £3m. HDL's financial history is summarised in the table 15.1.

Table 15.1
HDL's accounts 1959–72
(£000s)

Year	Income	Operating costs	Deficit (Surplus)* for year	Deficit carried forward
1959	nil	10.5	n.a.	11.0
1960	nil	250.3	268.1	279.1
1961	1.8	501.8	535.3	814.3
1962	46.8	569.2	607.9	1422.3
1963	13.0	219.0	336.6	1758.9
1964	211.0	185.3	134.0	1892.9
1965	52.0	209.0	336.1	2229.0
1966†	426.0	176.0	(108.3)	2120.6
1967	515.4	253.3	(53.7)	2066.9
1968	82.8	34.2	115.6	2182.5
1969	28.5	249.1	414.4	2597.0
1970	80.9	233.6	368.7	2965.7
1971	237.2	132.6	83.3	3049.0
1972	108.4	148.4	193.6	3246.2

* after charging depreciation
† 9 months only
Source: HDL's accounts

Operating costs include intramural expenditure and expenditure in industry. It is interesting to note that, although the Technical Group was closed down in 1967, HDL's operating costs are still running at well over £100,000 p.a. By 1972 cumulative gross operating expenditure (including depreciation) amounted to £5m. Total expenditure by NRDC, including the direct loan of £1m. to BHC to finance the SRN4, amounts therefore to just over £6m. (excluding expenditure by Tracked Hovercraft Ltd and the former shareholdings of NRDC in BHC).

Financial assistance has been given to industry in three main forms. Firstly, HDL has paid for the design and production of some craft (SRN1, VA1) entirely, and owned all the rights in them. Secondly, it has shared development costs of craft with industry, usually on a 50–50 basis. The SRN2, VA2, VA3, D1 and 'GERM' were all financed in this way. These craft were not seen as production craft although a provision was made in the agreements that, if sales did occur, HDL would receive some payment. In the case of the first three craft mentioned above, the firms concerned bought out HDL's interest. In both the first and second types of finance, the decision to invest was made principally on technical rather than on economic grounds. The third type of assistance was given solely on a recovery basis. CC4, CC5, VT1, AV2 and HM2 were assisted on these terms. (In the case of the SRN4, the NRDC provided a loan (£1m.) directly to BHC. This special arrangement was made because of the existence of NRDC's shareholding in BHC and of the relatively large sum involved.) Repayment of loans is on the basis of future sales. For example, HDL's advance of £16,000 to Hovermarine Transport for the HM2 was repayable on the sales of the third to the tenth craft at a rate of £2400 per craft. Repayment is not limited to sales of the particular craft type originally assisted: a levy can be made on any 'lineal descendant' of such craft. For example, HDL regards the BH7 as being a derivative of the SRN4. Because the SRN5 and 6 would probably have been regarded as a descendant of the SRN2, the decision of Westland to buy out HDL's share in the latter meant that no levy on sales was payable. Without access to the terms of agreement, however, it is difficulty to say whether this resulted in financial gain for Westland. Financial assistance of various kinds has therefore been provided by HDL and has thus given it considerable flexibility.

HDL's income has consisted of two main elements. Firstly, it has received a levy on sales due in return for its loan finance. Secondly it has received a licence income on its patents. This income consists of a non-recoverable down payment in advance of royalties and a levy of between two and five per cent on all future sales for the duration of the agreement.

There are several aspects of HDL's licensing policy that are worthy of note. Firstly, it has endeavoured to satisfy itself that the applicant has a craft that is technically sound and that he has the capacity and resources to produce and market the craft. In the House of Lords in 1968, a government spokesman defended some delay in the granting of licences to British applicants on these grounds.

> ... as regards licensing in this country the NRDC has a dual function. Not only does it license in general terms but it also has to ensure that a sensible industry grows up. This means that it uses its licensing policy to ensure that firms which do not have the prospects of making a go of it in this difficult new industry do not get in on the market and take up some of the not limitless resources of skill and know how.[1]

In retrospect, it is clear that HDL has been largely unsuccessful in identifying applicants with the right qualities. This lack of success may be largely inevitable and, more fundamentally, may reflect the economically marginal nature of commercial hovercraft operation. However, it does seem that in certain cases, especially with the smaller company, the provision of finance *alone* is insufficient; detailed managerial help may also be necessary. With strong management assistance, Hovermarine, for example, might have avoided liquidation. NRDC is reluctant to engage in this kind of activity because of the added strain to its own resources. However, management assistance seems to be as much a need of small innovating companies as finance. Apart from BHC, in which it terminated its interests in 1972, NRDC has had no equity holdings in any other company in the industry.

A second feature of HDL licensing is that a licensee is given access to all HDL's patents (and future patents). This arrangement means that HDL can utilise the life of its latest patent as the term of the agreement. However the effectiveness of such an agreement will depend on how far licensees can successfully argue that they are not using these latest patents. Since the key patents occurred in the early 1960s and HDL's own technical effort has now virtually ceased, it is doubtful whether the agreements currently in force could be sustained beyond 1980 at the latest. It is unlikely that HDL will have broken even by that date (although levies to repay the development loans would still continue to be payable).

Licence agreements are not standardised. HDL has from time to time imposed restrictions on the types of craft that can be produced by a licensee, and on the markets that can be attacked, tending usually to narrow down an applicant's objectives in the first instance. The down payment is determined on a rule of thumb basis, usually being the amount of royalties that would become payable in the fourth half-year after the craft has started to sell, using the applicant's own sales estimates.

Using a purely financial yardstick, HDL's involvement in hovercraft has so far been singularly unsuccessful. Its financial failure would be even more marked if the advance on royalties which never in fact materialised were excluded, and if account were taken of the timing of expenditure and receipts.

A number of possible reasons for the present financial position of HDL may be suggested. Firstly, it may be argued that insufficient control was maintained over HDL's own expenditure and that commercially unattractive avenues of R & D were explored without due regard for their likely financial return. One way in which the validity of this argument in relation to HDL's intramural expenditure may be explored is by looking at the numbers of those patents that have been allowed to lapse. HDL is likely to let a patent lapse only if it seems that the patent is unlikely to be commercially exploited and thereby to earn royalties. By September 1973, thirty HDL patents had lapsed (table 15.2). There is some evidence that HDL has become more stringent in its 'weeding out' process: of the last ten patents allowed to lapse, none were renewed beyond the ninth year, whereas the remaining twenty went beyond the ninth year before

Table 15.2
HDL: lapsed patents (at 1 September 1973)

Last year in which renewal fees were paid	No. of patents
6th	1
7th	6
8th	1
9th	2
10th	4
11th	7
12th	3
13th	5
Patents not renewed after fourth year	1
Total	30

Source: Patent Office records

Table 15.3
HDL's existing patent portfolio: age of patents*

Age of patent (years)	No. of patents
14	2
13	10
12	7
11	5
10	14
9	11
8	21
7	18
6	13
Total	101

* Includes all HDL patents where the complete specification was taken out up to and including 1967.
Source: Patent Office records

lapsing. Of course, many of the later patents taken out by HDL are still fairly young, and therefore many of them may still be allowed to lapse before the full sixteen-year term is completed. Table 15.3 gives a breakdown of the age of the existing patents, the complete specifications for which were published up to and including 1967. In the absence of comparable data it is difficult to assess this information. However, for the first three years of the sixties, about one-third of the patents taken out (thirteen out of forty) have been allowed to lapse. This seems a high percentage, especially as anticipated royalties would only have to be in excess of renewal fees for an economic case to be made for keeping a patent in force, *once a patent has been granted*. It is easy of course to be critical with hindsight. (The lapsing of the patents may be defended on the grounds that the relevant patents did not improve HDL's patent position. This then raises the question of why they were taken out in the first place.) It would however be of value to know on what basis HDL allocated its resources to different projects, and on what basis and by whom its work was reviewed.

Secondly, and linked with the first point, it is clear that the market did not develop as originally envisaged. Without access to the information available to HDL in the early 1960s, it is again impossible to know whether this could have been foreseen. Industry had certainly shown little interest in the hovercraft in the early days; unfortunately, however, industry's interest in a radically new concept is not an invariably good guide to its eventual commercial viability. The view that industry's judgement might be wrong was probably reinforced by the fact that the hovercraft's hybrid nature did not fit in easily to the technologies of the aircraft, shipbuilding or motor industries. Very few firms are involved in more than one of these industries. Consequently, rejection could be expected. This of itself does not of course make NRDC's decision to support hovercraft a correct one. HDL made its own evaluative studies of the commercial potential of hovercraft at different stages of its life, on which in part it presumably based its work. There is however one drawback to such internal studies that may be relevant here: there is a strong incentive for *any* organisation to provide evidence that justifies its own continued existence and expansion. This may tend to colour its estimates and calculations. To some extent this may have been true of HDL's evaluative work, although it should be noted that it was part of its policy to emphasise to new recruits that the life of the Technical Group was likely to be limited. There is however a strong case for arguing that, periodically at least, studies of this nature should be made or reviewed by an independent, uncommitted body.

Thirdly, the licence terms may have been negotiated at too low a level. It is unlikely however that there was much room for manoeuvre in this respect: average royalty rates received by the sample of firms chosen in the study by Taylor and Silberston (p. 45) was about four per cent of the net selling value of output. Taylor and Silberston's study also raises a more fundamental issue. This study showed (p. 173) that very few UK firms license other UK firms to work their patents. This seems to suggest that firms prefer to recoup their outlays on R & D, etc, from the profits on their own production, rather than from royalty income, presumably because the latter is less lucrative. A body such as HDL or NRDC, which has to rely almost solely on royalty income to recoup its own development expenditures, may therefore be operating under a financial disadvantage (compared with private industry).

Fourthly, some might argue that HDL's Technical Group was wound up too soon, that major breakthrough was just around the corner. This seems unlikely in view of subsequent events.

Any assessment of the overall effect of HDL on the development of hovercraft in this country must necessarily be speculative, since how the industry would have developed both technically and commercially *without* HDL cannot be known for certain. However it is reasonably clear that a prototype would not have been built in 1959 had HDL not provided the funds. The Saunders-Roe Board had already rejected a similar private venture proposal because of lack of finance, and no other firm had expressed interest. This rejection might have been an automatic reaction of a company that relied heavily on government funds for its income; on the other hand, Saunders-Roe had already undertaken a number of private venture projects, which is an indication of its willingness to operate on this basis. Interest in hovercraft from both industry and (particularly) the defence departments intensified very considerably after the SRN1 launching.

Whether the other craft types produced by industry with aid from HDL (and NRDC in the case of the SRN4) would have been forthcoming with the same or different designs is much more difficult to assess, and is tied up with the whole question of the commercial viability of hovercraft. Bearing in mind Westland's conservative R & D policy and its profitability in other spheres, and Vicker's shortage of financial resources and lack of management interest in hovercraft, absence of public aid may well have resulted in fewer craft types being produced. Again however it may be significant that the most successful line of craft to date, the SRN5/6, was a private venture financed by Westland. (The SRN5 was only a quarter of the SRN2's weight indicating a very substantial change in Westland's attitude towards the economics of hovercraft operation.) CC1 and CC2, D2, SH2 and EM2 (and a number of 'fun' hovercraft) were also private ventures. However, the technology embodied in these craft owed a considerable amount to earlier craft assisted by HDL.

HDL's own contribution to hovercraft technology as measured by patents has been mentioned in an earlier section. On the more general level, several people in industry and elsewhere commented in interview that HDL's preoccupation with its own patent position made it secretive and inward-looking. Westland's relationship with HDL's Technical Group also seems to have been a distant one, generated again by the relative patent position of the two organisations. Westland's acquisition of the Latimer Needham patent (p. 255) made the company somewhat less dependent on HDL's technology. Since Westland was the major producer of British hovercraft, the *raison d'être* of HDL as a centre for hovercraft expertise was therefore open to challenge. On the positive side however, apart from its technical contribution, HDL was acting as a training ground for hovercraft engineers and designers. Hovermarine and ACE were founded by ex-HDL employees; Vosper Thornycroft obtained its first manager for its hovercraft activities from HDL. All three companies have used HDL skirt technologies extensively. Several HDL ex employees have gone to the USA and Canada.

One final comment must be made on NRDC's financial commitment to hovercraft : it must be seen in the light of NRDC's *total* activities and of its objectives (see p. 144). By the very nature of its work, some individual projects are inevitably going to show losses.

15.2 Government departments

The involvement of government departments in hovercraft development has been in two main forms: research and evaluation activities, and 'straight' purchase of hovercraft for military activities. Available estimates of the total expenditure by government departments on hovercraft work are rather impressionistic and vary according to the definition of 'expenditure' used. One estimate given to the author was that total R & D expenditure of all kinds, including intramural work, purchase of craft for evaluation purposes and NRDC expenditure, by the end of 1974 amounted to approximately £15m. Thus about £10m. has been spent by government departments alone. Unfortunately, no breakdown by year was provided.

The research and evaluation activities have been carried out in three main forms: firstly, through small programmes of research at different state research establishments such as the Royal Aircraft Establishments at Bedford and Cardington and at NPL (although at the time of writing these programmes are being run down); secondly, through the Inter-Services Hovercraft Unit (IHU) at Lee-on-Solent. The IHU, which was set up in 1962, is administered by a committee which takes 'inputs' from D(T)I and the Ministry of Defence. The latter is principally concerned with operational requirements while the former's interest is more in terms of technical considerations. The IHU has evaluated the CC2, VA2, CC7, Hovercat, SH2, AV2 and to a lesser extent the VT1, and has acquired through its sponsoring ministries an SRN3, a BH7 and various SRN5/6 craft. The work of the IHU does of course have a useful feedback to industry since, within the limits of the Official Secrets Act, the producing firm is provided with a trials evaluation report without charge. The IHU does not evaluate civil craft, although it is difficult to tell *ex ante* whether a craft intended principally for the civil market has any military applications.

The third channel for government research work is in industry itself. After the contraction of HDL in 1967, the Department of (Trade and) Industry assumed principal responsibility, through its Hovercraft Branch, for placing contracts in industry. The initiative for formulating what work is to be done under these contracts has been primarily with industry. Contracts have been given to most of the firms in the industry as well as to engine manufacturers (especially Rolls Royce) and suppliers of skirt materials, although the lion's share has gone to BHC. The last installment of government money, after which no further special assistance will be granted, was announced in March 1972. Half a million pounds was to be spent intramurally and £1½m. in industry on development work. The latter was to be on projects devoted to improving existing craft, rather than on entirely new concepts or designs; for example, BHC has produced a twin-propeller version of the SRN6. No provision is made for the repayment of this money via a levy on sales, although the contracts were on a shared-cost basis. BHC, Vosper Thornycroft and Hovermarine Transport have all received funds under this last programme, although the bulk of the money has again gone to BHC.

The government's decision to withdraw public support from the hovercraft industry after the latest round of contracts was based on the study (referred to earlier) undertaken by the Programme Analysis Unit, which concluded that, as far as civilian high-speed marine craft (which includes hydrofoils) were concerned, 'the benefits from increased sales resulting from further government support of R & D etc, are . . . somewhat limited'.[2] The study did not however concern itself with military or industrial applications, two markets which look somewhat more promising (but see below).

The basis on which the figure of £2m. for the latest round of expenditure was arrived at and on which the sum was divided between different organisations is not known, although it is likely that the distribution of the funds was decided principally on the perceived technical merits of the proposals put to the Branch. As far as the author knows, no estimates of even a very crude nature were made of the likely economic benefits from the improvements to be carried out under the D(T)I contracts. In view of the fact that the money was intended for developing existing craft, rather than for the production of completely new craft, the calculation of the likely benefit would presumably have been much easier.

D(T)I (or its predecessor, the Ministry of Technology) also bought craft for its own evaluation work. It purchased the prototype CC7 and also an HM2. Similarly, the Ministry of Aviation purchased the first CC2 and RAE and the Ministry of Defence, the second CC2.

Purchases of hovercraft for defence purpose have been mainly for the 200 Squadron Royal Corps of Transport, formed in 1966. At the end of 1972 the Squadron had six SRN5/6s. For the producers government purchases provide not only revenue, but also an element of acceptability which may be useful in procuring other sales. For example, the establishment of the Squadron was probably responsible in part for the SRN6 orders received from the Iranian Navy. The government did not buy any of the Vickers craft. One former Vickers employee argued in an interview that, had it done so, the company's interest might have increased rather than waned. Government purchases in the early stages of a new development may also lower costs and thereby improve commercial prospects. However, the Army is now (1974) closing down 200 Squadron and this is bound to have some repercussions on hovercraft sales for military purposes. The reason given for the closure – that 'hovercraft will not make a contribution commensurate with its cost in the areas in which the Army is principally interested'[3] – suggests that even in the military sphere applications will be limited to specialist uses.

The government has also provided other forms of assistance. For example, it has supported a service on the Thames run by International Hoverservices. It has also provided limited financial assistance for export promotions.

NOTES on Chapter 15

1. *Hansard* (Lords) Col. 982 (26 March 1968)
2. K. M. Hill 'A Note on Some Economic Considerations for Government Policy towards Hovercraft and Hydrofoil' *Hoveringcraft and Hydrofoil* (December 1970)
3. Quoted in *Hoverfoil News* (14 February 1974)

16 Conclusions on the Case Study

As mentioned in the Introduction to this part, it is still too early to say whether in economic terms hovercraft development as a whole in this country can be regarded as successful from an *ex ante* viewpoint. On a social cost–benefit level, the account to date is still in substantial deficit. On the private side there are some encouraging signs. BHC, Hovermarine Transport, Air Vehicles and ACE have earned positive rates of return in the last few years, although BHC's *average* rate of return since 1967 has been very low. Furthermore, it must be remembered that Air Vehicles' substantial profits have not come from sales of its craft. ACE's record was very good in 1971 and 1972, but its performance fell off dramatically in 1973. Enfield Marine and Vosper Thornycroft have yet to sell any of their craft. On the operating side, neither of the two big operators had produced a positive rate of return by 1972 despite the fact that they have quickly established themselves as major carriers on the cross-Channel services, particularly in the French Straits. Hovertravel has been profitable, but the results for 1972 have shown its extreme vulnerability as a small company to changing market conditions. On both the manufacturing and operating sides, several companies have either withdrawn, been taken over or gone into liquidation. NRDC's financial involvement via HDL has so far been financially unrewarding.

The impression gained from this record of the industry's development to date is not therefore one of outstanding success. The picture may of course change – it is still early days, at least for a major invention – and the author looks forward to monitoring future developments.

It seems clear that one reason for this record is that the craft so far developed do not appear to have been sufficiently competitive to provide a profitable alternative means of transport in the commercial ferry market. It is mainly for this reason that both Vosper Thornycroft and BHC have turned from the civil to the military and paramilitary market where cost considerations may be of less importance. (It is ironic that the Defence Departments rejected hovercraft in 1958 and that, as a result, NRDC was inevitably forced to concentrate on civil applications if it was to become involved at all.) Neither Sealand nor Air Vehicles, two fairly new firms, are aiming at the commercial ferry market.

It can of course always be argued that if BHC had adopted a somewhat different line of development prospects in the commercial market might have been much better. Clearly, the technology this company utilised was drawn principally from aircraft manufacture and its staff were used to working on projects of a technologically advanced nature which in part, at least, were geared to military needs. This expertise *may* have been somewhat inappropriate for the full exploitation of the air cushion principle – at least in the commercial ferry market. It is however too early too make a firm judgement on this matter. In this respect it will be interesting to watch the progress of Hovermarine Transport, whose own technological background is much less committed and whose end-product, which is aimed at the commercial market, is rather less sophisticated.* The BHC craft have however sold relatively well in the non-commercial market where the company has achieved substantial export sales. This may be the area in which the amphibious hovercraft will eventually make its major impact.

As the industry develops and changes, the inventor who started the ball rolling continues to work independently. Cockerell was an 'outsider' with virtually no previous knowledge of anything that might be even vaguely connected with the air cushion. His experience has shown that full patent protection for an invention may be beyond the reach of the inventor with limited resources. It has also shown that an inventor cannot hope to keep control over the way in which his invention is exploited, where that invention requires substantial resources for its development. Nor would this

---

* While this book was in press, it was announced (*The Financial Times*, 29 January 1975) that Hovermarine Transport is to develop its HM5 design with the aid of £½m loan from NRDC. (The loan will be repaid by a levy on sales). This is significant in the context of the discussion in the text.

necessarily be optimal : the existence of inventive ability does not necessarily imply innovative ability. Indeed an inventor's personal commitment to the product of his own mind and hands may strongly influence his perception of its commercial and technical potential.

The concept of the 'air cushion' as developed by Cockerell was not entirely virgin territory : work had already been done by others on the possibility of using the air cushion in a variety of roles. However, Cockerell was the first to get his work taken up on a large scale. His success it seems was due to a host of factors : the technical superiority of his ideas, his own persistence and enthusiasm, the interest of the (then) Ministry of Supply, etc. These factors combined to provide Cockerell with his opportunity. Industry's initial refusal to take up his work will have to be judged in the light of the final outcome of hovercraft development.

The technical development of the industry since 1959 has reinforced the view that the inventor working independently or in a non-commercial organisation, often with little direct experience in the field in which he invents, is still an important source of ideas. BHC's contribution to the hovercraft's technical development has been very considearble – especially in relation to the continuous improvement of materials, designs, etc, but it was an independent consulting engineer who gave them their basic patent on the skirt. The other basic skirt patent came from Bliss at HDL, an organisation which, in its early days at least, had a fair amount of freedom to organise its own programme of work. The technical history of the industry has also shown that most development has been in the form of small but steady accretions to knowledge by numerous organisations rather than isolated 'flashes of genius' by a few individuals. Even the basic patents must be seen in the context of preceding and following work without which they would have been virtually useless.

Competition among producers (including HDL) has probably had a beneficial effect on the pace of technological development. Westland and Vickers were technical rivals in the early 1960s and as a result developed their own individual skirt systems, as did HDL. Since the mid-1960s, the entry of new firms has also meant greater experimentation with new designs, materials, etc.

The effect of government involvement on technical development is very difficult to determine. As far as inventive activity as measured by patents is concerned, HDL does not appear to have acted as a stimulus to the industry. Nevertheless, HDL has clearly made an important contribution to hovercraft technology both intra- and extramurally, much of it of the 'how not to do it' type. Most of the craft types produced to date have been assisted by HDL (or the D(T)I). However, many of HDL's patents have now been allowed to lapse. And it has already been noted that the SRN5(6) – BHC's most successful craft – was produced without direct government backing. It will be interesting to see what happens to Sealand's and Enfield Marine's craft – both privately sponsored projects.

It was orginially NRDC's intention that the manufacturing side of the industry should develop in an orderly and 'sensible' fashion. It was not long however before this plan became irrelevant as some companies began to withdraw and others to enter. The failure and withdrawal of companies during the last fourteen years has been due to a variety of reasons and can only be fully understood in the context of particular circumstances. Denny had a long record of technical innovation but its hovercraft work could not survive the decline of the company's ferry market. Hawker Siddeley was wrapped up in its aircraft interests. Hovermarine over-reached itself mainly because of poor financial control. Vickers had a good technical record at Supermarine but its main management was unenthusiastic. Hoverair's markets did not materialise as anticipated.

It is futile to attempt to find a common thread through these events since each firm operated under different conditions and within different environments. However, they all probably underestimated the costs and time involved in developing a new technology. Viability may of course never have been achieved; on the other hand, it may have been 'just around the corner'. For example, both Denny and Hovermarine might have achieved eventual success in the hands of the original companies, had finance been forthcoming.

Several new companies have been formed to produce hovercraft and associated equipment. A variety of reasons have again been present although for three companies at least the frustration of the founders in their previous employment was an important factor in their companies' formation. The enthusiasm and total commitment of company founders for their own ideas was very evident in interviews carried out in this study (several had never been in business before). Unfortunately, this has not always been accompanied by a realistic assessment of costs or markets or of the complexity of the technology involved. The failure of Hovermarine in 1969 did however make new companies somewhat more cautious :

both Air Vehicles and ACE have attempted to maintain closer control over their costs.

On the operating side, too, numerous companies have been formed to try out hovercraft on different routes. Most have failed, again through poor assessment of the costs involved and of markets. The experience of British Rail and Hoverlloyd has shown the substantial costs involved in implementing a new service, and the basic question re-emerges : Are these losses due merely to teething troubles, or do they reflect more basic economic disadvantages of hovercraft operation?

Established companies have also embarked on hovercraft production and operation. The motivating factors behind their entry have differed. The first four producing firms entered for defensive reasons: because of an actual or potential decline in their existing business. Hoverlloyd set up its hovercraft services because it was effectively banned from operating conventional ferries. Britten-Norman and Vosper Thornycroft (along with the new companies), however, were 'active' innovators, becoming involved for more positive reasons. A constant source of new entrants has ensured that new ideas and possible markets are tested; there can be little doubt that the industry as a whole has gained in this way although the private costs have often been substantial. Air Vehicles, Sealand and ACE are manufaturing products aimed at markets that BHC has not attempted to exploit. On the operating side, Hovertravel's work is an outstanding example of a small company exploring hitherto untapped fields. The record shows, too, that profits for small new companies can be high; but their vulnerability to changes in the market is also very much greater. The smaller company is often more flexible than its larger rival and there may be a greater incentive to succeed because more is at stake, but it may lack basic managerial skills and access to financial resources.

As far as finance is concerned it may *not* be the complete policy answer to provide new avenues of possible assistance; this may remove some of the very spurs that maintain the small company's ingenuity and entrepreneurial skill. There may therefore be some inevitability (however much disliked by the small firm) in having a division of labour in which a small adventurous company undertakes the basic work and provides an important marketing or technical idea, and is then taken over by a larger, financially stronger, company.

While patents may appear at first sight to have played an important role in the industry's development, the existence of HDL as a quasi-government body holding some of the key patents has meant that entry has not been very difficult, although it may have been delayed. The picture might have been different if HDL had been an entirely private body. Patent specifications are however only one aspect of technical know-how. The industry has shown (to its cost) that the speed at which a company can enter the industry will depend on its knowledge of a much wider concept of hovercraft technology. The new entrant may build this knowledge up internally, or may 'buy in' key personnel. Whichever strategy is adopted, costs of acquiring the technology may be substantial.

Being a government department or a public body does not necessarily reduce the difficulties of spotting a potentially successful invention, although it does decrease the penalties for failure. In the light of its cumulative deficit, NRDC's involvement in hovercraft may be criticised with hindsight, although without access to the information available to it at different times in the past fourteen years, we cannot be dogmatic over whether its decision-making processes could have been improved. It is possible however that the hovercraft project gathered its own 'internal momentum' within NRDC and thereby generated a strong technical lobby to justify continued support. (In the early days of course, with so many uncertainties, it would have been possible to argue a strong case either way.)

The changeover from NRDC to MinTech and then DTI also raises the basic question of why the constraints under which NRDC operated were not a suitable framework for the continued support of the hovercraft in industry. If financial assistance to industry could not be justified under the NRDC's terms of reference, under what terms *could* it be justified? While there may be a strong case for assistance from government departments (with its somewhat looser constraints) in the early phases of a development programme, it is more difficult to make a firm case for financing later stages, when the degree of technical and commercial uncertainty has been reduced. It should be remembered too that the latest grants were for *improvements* to existing craft, and not for entirely new projects where the uncertainty may still be considerable.

Government involvement via either Departments or the NRDC has clearly been an important facet of the industry's development.

No figures have been publishel on the percentage of total R & D expenditure financed from public sources, on a loan or grant basis, but it may be of the order of between fifty and seventy-five per cent. For some small companies, eg Air Vehicles, public funds have been crucial to their existence. As already indicated in the discussion of HDL's role, it is difficult to assess the precise effect of this government expenditure without access to more data. Whether it has meant, for example, that companies have undertaken work under a contract that would otherwise have been privately financed or that would *not* otherwise have been undertaken because it was not considered profitable is uncertain. Following through the results of individual contracts and examining the extent to which they are used in production would be of considerable help here.

Much of this study has necessarily been highly speculative. Developments over the next few years, both in this country and abroad, should go a long way towards giving a clearer picture, and providing longer runs of data on both technical and economic matters.

APPENDIX 1

Time Savings (to 1972) from Hovercraft Operation

Passenger data on all the main hovercraft routes for each year of operation were first collected. These data are shown in table A1.1. They vary a little in the degree of refinement but the figures provide a fairly accurate picture of passenger movements.

Time savings were calculated on the basis of a comparison between hovercraft journey times and journey times of the nearest alternative transport. These savings were calculated as follows:

Table A1.1
Passengers carried on principal hovercraft routes, 1965–72
(000 passengers)

Year	Southampton – Cowes	Portsmouth – Cowes	Portsmouth – Ryde	Southsea – Ryde	Cross-Channel
1965				180.0	
1966	38.0			349.0	
1967	95.0	45.1		286.0	
1968	127.0	42.6	101.3	341.0	27.0
1969	167.9	27.4	228.4	377.0	420.5
1970	176.0		152.4	352.0	853.5
1971	191.2		257.2	389.0	1203.5
1972	186.8		86.7	400.0*	1258.1

* estimate
Source: Operating companies

Cross-Channel	1 hour
Portsmouth–Ryde	
Southea–Ryde	20 minutes
Portsmouth–Cowes	
Southampton–Cowes	30 minutes

318 APPENDIX 1

These estimates are open to some criticism. Firstly, rarely do the terminal points of the alternative services coincide with those of the hovercraft services (see p. 294). Secondly, there may be additional time savings to be made from travelling by hovercraft that are not estimated here, eg less time spent queuing for boats, easier car parking, etc. Neither of these criticisms however is likely to be very significant. Thirdly, the Portsmouth–Cowes route does not have even a rough counterpart in an alternative means of transport. This route is so unimportant in quantitative terms, however, that it has been treated in the same way as the other services starting at Portsmouth.

A division between working and leisure time saved was made on the assumption that seventy-five per cent of travellers on the Southampton–Cowes route were travelling in working time, while the figure was put at twenty-five per cent for all other routes. This assumption is probably highly favourable to hovercraft in the valuation of time saved.

Two discount rates – ten and five per cent – were used to bring the figures to 1959. The results are shown in table A1.2.

Table A1.2
Time savings to date as at 1959
(000 hours)

Discount rate	Working time saved	Leisure time saved
10%	501	1182
5%	909	2164

Valuing such time creates considerable conceptual and empirical problems and has been a topic of much controversy among economists working in the cost–benefit field.[1] Foster and Beesley, in their study of the Victoria Line, relying on an earlier investigation, valued working time in 1962 at 36p per hour and leisure time at 25p.[2] For 1959, of course, the figures would be correspondingly less. However, in order to bias the figures in favour of hovercraft, the somewhat arbitrary figures of 50p and 25p respectively have been selected. These values give the results shown in table A1.3.

There are three additional factors that must be borne in mind when interpreting these data. Firstly, the passenger figures in table A1.1 include a substantial number of children; thus the figures will tend to overstate the value of the savings made (assuming that

APPENDIX 1 319

Table A1.3
Value of time saved to date
(£000s)

Discount rate	Working time saved	Leisure time saved	Total
10%	251	296	547
5%	455	541	996

children's leisure is less valuable than that of adults!). Secondly, a proportion of those travelling will be foreigners and hence their time savings should not strictly be included in the calculation of national benefits (such travellers do however provide foreign earnings). Thirdly, a proportion of the passenger traffic will be *generated* by the existence of the hovercraft. Time savings are an inappropriate measure of the benefits gained by the passengers concerned. Unfortunately, data were not available to refine the figures in the light of these qualifications.

NOTES on Appendix 1

1. See for example, P. Watson and N. Mansfield 'The Valuation of Time in Cost Benefit Studies' in J. N. Wolfe *Cost Benefit and Cost Effectiveness* London, Allen and Unwin (1973)
2. C. D. Foster and M. Beesley 'Estimating the Social Benefit of Constructing an Underground Railway in London' *Journal of the Royal Statistical Society* (1963)

APPENDIX 2

Total R & D Production and Operating Costs (to 1972) of Hovercraft Development

The estimates of costs given in table A2.1 are only intended to give an idea of the likely magnitudes involved. The figures have been built up from company accounts, press reports and data provided by companies. On the R & D and production sides they include expenditure by HDL, BHC (formerly Westland), Vickers, Denny, Hovermarine and later Hovermarine Transport, Cushioncraft, Vosper and Rolls Royce. In many cases arbitrary allocation to individual years of single expenditure estimates has been necessary. Foreign sales have been deducted from the figures. On the operating side, the estimates relate to expenditure by British Rail Hovercraft, Hoverlloyd and Hovertravel.

It is clear that the figures do not include all R & D, production and operating costs involved in the exploitation of hovercraft. All the activities of the armed forces and intramural work of govern-

Table A2.1
Estimates of R & D production and operating costs

Year	£m.	Year	£m.
1959	0.2	1966	2.1
1960	0.8	1967	2.9
1961	1.4	1968	6.3
1962	1.4	1969	12.6
1963	1.1	1970	10.8
1964	1.1	1971	15.8
1965	2.1	1972	14.1

ment departments are excluded, as are those of small producers and operators.

The figures in table A2.1 were then discounted at 15 and 10 per cent (the discount rates used here are higher than those in Appendix 1 to allow for the effects of inflation). The results are shown in table A2.2.

Table A2.2
Total costs discounted to 1959

Discount rate	Total costs in 1959 (£m.)
15%	16.1
10%	26.5

Further Work

It should by now be clear that the process of technological change is an extremely complex phenomenon and that, although some useful insights into this process – not least the fact that industrial R & D may be subject to economic analysis in much the same way as other activities – have already been gained by economists and others, much remains to be done. None of the fields discussed in this book is by any means exhausted : future research could usefully be done in such relatively well researched areas as the relationship between firm size and innovation. For example, while there is a rapidly growing literature on the effects of mergers – clearly an important influence on firm size – as measured by various financial indicators, particularly profitability, there is very little on their effects on the inventive and innovative record of the companies concerned.

However, in the author's view, there are two areas in particular that require (and would repay) much more effort. Firstly, although the government is responsible for over fifty per cent of all R & D expenditure in the UK (p. 116), very little is known about the decision-making processes that lead to the allocation of this expenditure between different uses, or about its impact – whether it is spent extra- or intramurally. A start might be made in this field by a series of case studies of government R & D contracts given to a particular industry. A prerequisite for such studies however is that the basic data are forthcoming not only from the companies concerned but also from the relevant government departments.

The second area that has been neglected in the study of technological change is the *location* of innovation and factors that affect it.[1] This is of course particularly relevant in the context of regional policy. The ultimate objective of this policy is presumably its own demise : to achieve a situation in which regional disparities still exist but where they are not politically unacceptable. If this basic objective is accepted, then the long-term policy question hinges on *how* overall regional decline as experienced in the past can be avoided without recourse to government subsidies. In the short term, the basic policy problem is how to ensure that industries *already* attracted by regional incentives can be induced to maintain and/or increase their employment. Both the long-and short-term questions are closely related to the location of innovation. If industry can undertake its own 'rejuvenation' in a particular area, introducing new products and processes as old ones disappear, then the need for regional assistance will also disappear. This rejuvenation may occur not only within the confines of existing firms, but also through the formation of 'spin-off' companies.

Most of the work on regional economics to date however has paid very little attention to these issues. We do not know, for example, whether *within* given industries employment expansion in a region is the result of growth in that region of the output of established products, possibly with low growth potential (with a corresponding decline in such output produced in other regions), or of *new* products, with high growth potential. The factors that exercise an important influence on the decisions made by existing firms – particularly those producing on a multi-regional basis – on where to site initial production of their new products have not yet been identified. Again, little is known about the formation of spin-off companies, the initial problems they face, the types of firms and industries that are likely to be most fertile in 'fathering' such companies and the extent to which such companies are likely to be located near to their parent. Some work on spin-off has been done in the United States;[2] this has shown the importance of the phenomenon for regional development.[3] Relatively little work however has been done in this field in the UK.

The two areas mentioned above seem particularly deserving of study. This is not to deny, however, that other gaps in our knowledge and understanding in this field also exist; at least this much should be apparent from the preceding pages.

NOTES on Further Work

1. Fortunately, the cupboard is not completely bare. See R. J. Buswell and E. W. Lewis 'The Geographical Distribution of Industrial Research Activity in the UK' *Regional Studies* (1970) and the reference contained

therein, and C. Freeman and A. Young *The Research and Development Effort in W. Europe, North America and the Soviet Union* Paris, OECD (1965) chapter IX.
2. See for example A. C. Cooper 'Spin-offs and Technical Entrenemship' *IEEE Transactions on Engineering Management* (1971) and E. B. Roberts's study cited on p. 101 and the references contained therein.
3. E. B. Roberts and H. A. Wainer 'New Enterprises on Route 128' *Science Journal* (1968)

Index

Abramovitz, M.   9, 36
Acts of skill   30–2
Acts of insight   30–2
Admiralty   246
Air Bearings Ltd   237, 242, 277
Air cushion
  early work on   251–3
  and flexible skirts   253–4
Air Cushion Equipment Ltd (ACE)   237, 271, 277, 278, 279, 305, 309, 312–3
Airline design, innovation in   65
Air Vehicles Ltd   237, 242, 255, 256, 270, 275–6, 278, 279, 280, 282, 284, 285, 309, 310, 312–3, 314
Allen, J. A.   225
Applied research   25, 168
Arrow, K.   50, 76–81, 90, 128, 129, 131, 154
  and Capron, W. M.   95, 107, 111, 112
Atomic Energy Authority   119, 122

Bain, J.   88, 91
Barnaby, K. C.   243
Basic research
  defined   21
  economic characteristics of   22–5, 124
  industrial expenditure on   168
  in universities   215–6
Beardsley, Colonel   252
Becker, G.   106, 112
Bevan, E. G.   225
Blair, J. M.   48, 50, 174, 189
Blank, D. and Stigler, G.   94–5, 111, 154
Bliss, D.   255, 311
Blume, S. S. and Sinclair, R.   220, 221, 225

Bolton Committee on Small Firms   100, 285
Bosworth, D. L.   34, 49
Brain drain   108
Bright, A. A. and Maclaurin, W. R.   81, 90
British Hovercraft Corporation (BHC)   232, 236, 237, 239, 240, 242, 254, 255, 256, 257, 258, 259, 262, 270, 271, 276, 283, 287, 299, 300, 307, 309, 310–11, 313, 320
British Rail Hovercraft ('Seaspeed')   236, 241, 288–9, 290–1, 292, 294–5, 296, 313, 320
Brown, G.   71
Browning, W.   296
Britten-Norman Ltd   236, 270, 272, 273, 275, 278, 279, 287, 290
Buswell, R. J. and Lewis, E. W.   323
Buxton, A. J.   13, 27
Byatt, I. C. R. and Cohen, A. V.   217–8, 219, 225, 246

Caldecote, Lord   247, 250
Caledonian Steam Packet Co   291
Carter, C. F. and Williams, B. R.   16, 20, 27, 68, 85, 91, 162–3, 186, 188, 189, 208
Caves, R.   154
Central Policy Review Staff   122–3
Clyde Hover Ferries Ltd   291
Cockerell, Sir Christopher   31, 215, 250, 311–2
  and HDL   248–9
  approaches to industry by   246–7
  approaches to NRDC by   247
  background of   244
  early experiments by   244–5
Cole, J. and S.   225

325

Comanor, W.   70, 86, 91
Concentration ratios
  and technological change   84–6
Concorde   141–4
Cooper, A. C.   324
Corlett, W. J.   90
Cost-benefit analysis of R & D
    135–44
  individual studies of   136–41
  costs in   141–4
Council for Scientific Policy   218,
    220, 224, 225
Cox, J. G.   56, 70
Cushioncraft Ltd   236, 255, 256,
    273, 320
Cyert, R.
  and George, K. D.   75, 90
  and March, J. G.   90

Dawson, C.   243, 297
De Havilland   246
Demsetz, H.   78–90, 129, 154
Denison, E.   9, 11, 12, 16, 26, 27
Denny Bros., Wm.   233, 266, 267,
    268, 269, 272, 274, 278, 279,
    285, 312, 320
Department of (Trade and) Industry
    233
Development   21
Diffusion of innovation   173–187
  empirical measures of   182–5
  measurement of   175
  theoretical issues in   175–182
Downie, J.   17, 90, 189

Eckaus, R. S.   17, 27
Edwards, R. S.   208
Enfield Marine Ltd   242, 255, 273–
    4, 276, 278, 279, 309
English Electric Ltd   246, 247
Enos, J.   71, 112, 174, 189

Fabricant, S.   26
Fairley, P.   250
Flexible skirts   31, 253–6
Foster, C.   82, 90
Foster, C. D.   218, 225
  and Beesley, M.   318, 319
Freeman, C.   6, 26, 52, 66, 70, 71,
    160, 161, 188, 324

Galbraith, J. K.   15, 59, 62, 63, 70
Gifillan, S. C.   31, 49
  and Ogburn, W. F.   33

Government R & D expenditure
  and private R & D   132–4, 159,
    322
  customer-contractor relationship in
    122–3
  economic basis of   123–35
  extent of   115–7
  inefficiencies in   130–2
  risk aversion and   129–30
Grabowski, H.   75, 163–4, 188
Gregory, R. G. and James, D. W.
    189
Griffiths, E. W. H.   235
Griliches, Z.   24, 136–7, 182, 183,
    185, 189
  and Jorgenson, D. W.   13, 26, 28
Grossfield, K.   155
  and Heath, J. B.   140–1, 154
Grubel, H. G. and Scott, A. D.   112

Hamberg, D.   43, 60, 62, 70, 71,
    91, 189
Hawker Siddeley Ltd   266, 267, 268,
    278, 312
Heath, H. F. and Hethington, A. W.
    208
Henderson, P. D.   154
Hill, K. M.   243
Hitch, C. J.   130, 154
Horowitz, I.   85, 91
Hoverair Ltd   275, 278
Hovercraft
  across the Channel   238, 317
  and conventional ferries   295–6
  and hydrofoils   293–5
  and the Channel tunnel   231–2
  costs and benefits of   230–3, 317–
    21
  directional control of   256
  finance for development of   280–6
  foreign producers of   286
  for the civil market   239
  markets for   238–42
  materials for   257
  operations   237–8, 288–9
  operating costs of   231–2
  patents for   258–60
  production of   235
  reliability of   291–3
  technological spin-off from   262–3
  types of   260–2
Hovercraft Development Ltd (HDL)
    31, 241, 247–8, 254, 255, 258–
    60, 262, 266, 267, 272, 273,

276, 285, 286, 287, 300–2, 305, 313, 315, 320
Hoverlloyd Ltd   235, 237, 288–9, 290–1, 292, 295–6, 313, 320
Hovermarine Transport Ltd   236, 237, 242, 243, 255, 271, 274–5, 278, 279, 282–5, 288, 292, 305, 307, 309, 312, 320
Hovertravel Ltd   236, 237, 275, 284, 289, 290–1, 309, 320
Humber Hoverservices Ltd   291–2
Hybrid corn research   136–7, 185

Industrial Research Establishments   120
Industrial R & D expenditure
  breakdown by type of work   168
  concentration of   156
  in the individual firm   162–6
  nature of   167
  output growth and   161
  patents and   170
  profitability and   160–1
  research intensities and   158–62
Innovation
  channels for   64–5
  defined   19–20
  established firms and   65
  firm size and   64–8
  leaders in   186
  location of   322–3
  market structure and   73–87
  new companies and   64–5
  risky nature of   20
  successful vs. unsuccessful   68
Invention
  as an economic activity   32–9
  defined   18–9
  demand factors and   37
  firm size and   59–63
  incentive for   76–80
  its nature   29–31
  measurement of   33–5
  scientific discovery and   36
Inventor, background of   63

Jewkes, J.   16, 37, 50, 54, 59, 60, 70, 71, 131, 154, 224, 230, 234, 245, 250
Johnson, H. G.   28, 112
Johnson, P. S.   70, 154, 208
Jones, P. M. S.   154

Kennedy, C. and Thirlwall, A. P.   26
Keynes, J. M.   15, 27
Know-how, employer's rights in   104–6
Kuhn, T. S.   28
Kuznets, S.   36, 49

Landes, D. S.   5, 26, 189
Langrish, J.   68, 71, 99, 101, 112, 224, 225
Latimer-Needham, C. H.   255, 305
Layard, P. R. G.   111
Layton, C.   27
Leibenstein, H.   189
Leonard, W. N.   161, 188
Lethbridge, D. G.   297
Lloyd, B.   71
Lynn, A.   174

Machlup, F.   45, 50
Mackenzie, A. E. E.   225
Maclaurin, W. R.   27, 71, 90
Mc Neil, D.   208
Mansfield, E.   17, 27, 57, 67, 70, 71, 83, 164, 165, 166, 182, 185, 186, 187, 188, 189
Markham, J.   90
Mason, E.   90
Matthews, R. C. O.   12–3
Metcalfe, J.   182, 186, 189
Miller, R. and Sawers, D.   71, 82, 90
Minasian, J. R.   13–4, 27, 162, 166, 188, 189
Monopolies Commission   50, 82, 90, 91, 99, 155, 295–6, 297
Mountbatten, Lord   246, 250
Mueller, W.   25, 28, 61, 70

Nabseth, L. and Ray, G. F.   175, 189
Napier Ltd   246
Nath, S. K.   154
National Research Development Corporation   64, 101, 155
  activities of   147–9
  aims of   144–5, 146
  assistance to industry by   145
  financial record of   150–2
  hovercraft interests of   233, 239, 240, 247, 248, 249, 252, 262, 263, 265, 266, 267, 269, 270, 285, 298–305, 310–12, 314
  small firms and   145

Nelson, R. R.   23, 27, 28, 224, 225
Nicholson, R. L. R.   154

Ogburn, W. F.   49

Panic, M. and Close, R. E.   297
Passer, H. C.   65, 71
Patents
  as barriers to entry   88–9
  as measures of inventive activity   33–5
  as restrictions on technology transfer   104–5
  breakdown of, by industry   169–70
  R & D expenditure and   170
Patent system   39–48
  compulsory licences under   40–1
  economic arguments for   42–5
  empirical studies of   45–7
Peck, M.   27, 70, 96, 112, 188
  and Scherer, F. M.   154, 155
Phillips, A.   85–6, 91
Polanyi, M.   220, 225
Price, D. K.   27
Programmes Analysis Unit   241
Public Accounts Committee   143, 155

Qualified scientists and engineers
  definition of   92–3
  international movement of   108–10
  mobility of   97–108
  shortages of   93–6
  technicians and   96

Rapson, J.   296
Ray, G. F.   182, 186, 189
Red Funnel Company   293
Reekie, D.   53, 70
Research and development (R & D)
  commitment of OECD countries to   5–6
  concentration of industrial   51–2
  economic growth and   6–13
  firms' budgets for   163–5
  intensities   54–5
  its nature   124–5, 197–8
  growth of   1–2
  productivity of, and firm size   53–4
  productivity and   13–5
  profitability and   166
  royalties and   172
  social costs and benefits of   126–7
  the spectrum defined   20–1
  threshold level of   52–3
  uncertainty in   127–8
Research associations
  activities of   199–204
  contract research in   205
  economic case for   197–9
  financing of   207
  in individual industries   192–5
  in Europe   190–1
  membership of   195–6
  present scale in UK of   191–2
  size of   196–7
Research Councils   121, 123, 211
Residual   7–9
Restrictive agreements and innovation   83–4
Restrictive Trade Practices Court   84
Richardson, G. B.   112
Robbins, Lord   27
Roberts, E. B.   101, 102, 112
  and Wainer, H. A.   324
Rose, J.   189
Rosenberg, N.   38, 50
Royalties   169–73
Ruttan, V.   30, 49, 257

Salter, W. E. G.   16, 27, 176–82, 183, 189
Sanders, B. S.   49, 69
Sanderson, M.   26, 111, 224
SAPPHO Project   68
Saunders-Roe Ltd   241, 246, 247, 248, 251, 265, 266, 267, 278, 279, 304
Scherer, F. M.   28, 59–60, 69, 70, 71, 74, 86, 90, 91, 102, 112
Schmookler, J.   16, 17, 19, 20, 27, 33, 35, 36, 37, 38, 49, 70, 71, 165, 245
Schumpeter, J.   18, 19, 20, 27, 30, 58–9, 61, 67, 70, 72, 73, 74, 75, 85, 257
Science
  defined   18
  discoveries in and invention   215
Science Research Council   222
Sealand Hovercraft Ltd   237, 255, 256, 276, 282, 285, 310, 313
Select Committee on Science and Technology   121, 149, 153–4, 155, 250

Shaw, Ronald   246, 247
Sherwin, C. and Isenson, R. S.   224
Shute, Nevil   154
Silberston, A.   50
Silk, L. S.   26
Solo, R. A.   131, 154
Solow, R. M.   9, 26
Spin-off companies   101–4
Stanton-Jones, R.   243, 255
Stigler, G.   85–6, 91
Stubbs, P.   28
Sturmey, S. G.   71, 90, 91, 112, 234, 235
Swann, D. and Mc Lachlan, D. L.   27
Sweny, L. A.   243

Taylor, C. T. and Silberston, Z. A.   14, 27, 45–7, 50, 70, 169, 172, 173, 189, 304
Technological barriers to entry   87–9
Technological opportunity   86–7, 160
Technology
  defined   17
  on the hoof   97–108
Thomas, B.   112
Tilton, J.   102, 106, 112, 154, 161, 187, 188, 189
Tracked hovercraft   149–50, 262

Universities
  concentration of R & D in   221–2
  evaluation of R & D in   213–9
  industrial support for R & D in   211–2
  nature of R & D in   213
  output of R & D in   214–5
  R & D expenditure of   209
  R & D finance for   210
  R & D productivity in   220
Universities Grants Committee   211
Usher, A. P.   29–30, 32–3, 49
Usher, P. J.   242

Vickers Ltd   233, 240, 249, 254, 266, 267, 268, 269, 278, 286, 287, 305, 311, 320
Vosper Thornycroft Ltd   232, 233, 237, 239, 240, 241, 251, 255, 273, 274, 278, 279, 286, 287, 291, 305, 307, 310, 313, 320

Watson, W.   208
Weisbrod, B.   137–40, 154
Weiland, C.   247
Westland Aircraft Ltd   232, 240, 249, 254, 259, 266, 269, 273, 278, 282–3, 286, 305, 311
Whybrew, E. G.   112
Wilkinson, G. C. G. and Mace, J. D.   111
Williams, B. R.   26, 27, 52, 70
Worley, J. S.   57, 70

Yamey, B.   80, 90